Signals and Communication Technology

More information about this series at http://www.springer.com/series/4748

Vishram Mishra · Jimson Mathew
Chiew-Tong Lau

QoS and Energy Management in Cognitive Radio Network

Case Study Approach

Springer

Vishram Mishra
Nanyang Technological University
Singapore
Singapore

Chiew-Tong Lau
Nanyang Technological University
Singapore
Singapore

Jimson Mathew
Indian Institute of Technology Patna
Patna
India

ISSN 1860-4862 ISSN 1860-4870 (electronic)
Signals and Communication Technology
ISBN 978-3-319-45858-8 ISBN 978-3-319-45860-1 (eBook)
DOI 10.1007/978-3-319-45860-1

Library of Congress Control Number: 2016952004

Printed on acid-free paper

This Springer imprint is published by Springer Nature
The registered company is Springer International Publishing AG
The registered company address is: Gewerbestrasse 11, 6330 Cham, Switzerland

Preface

The approval of free usage of frequency spectrum via dynamic spectrum access methods on a non-interfering policy basis in various countries has made the cognitive radio technology a prominent candidate for wireless communication systems. Most of the wireless devices and their supporting infrastructure available today are capable of handling multimedia content delivery over wireless channels. Therefore, quality of service (QoS) is an important parameter for users experience while utilizing the services rendered on their wireless devices. For realization of cognitive radio-based dynamic spectrum access over conventional devices with conventional services (VoIP, Video, text messaging, Web page, etc.), QoS is an important factor that need to be provided to the wireless devices and applications. The QoS is broadly defined via two network parameters: throughput and latency. Similarly, energy management is also an important criteria for increasing the lifetime of portable wireless devices. Different aspects of a wireless network contribute toward varying network parameters and energy consumption, with each aspect having its own weight in contribution.

In this book, we discuss a QoS provisioning framework for cognitive radio network, which identifies and divides attributes of system into different modules responsible for quality of service and energy management. These modules are independently studied and a relation between them is shown to depict the interaction between these modules. These modules are responsible for channel selection, media access control, self-co-existence, and energy saving. A relationship is drawn among these modules for a coherent decision followed by the system. The framework supports four access categories for service differentiation. The framework is also designed in such a way that it separates main communication protocol stack from quality of service engine and thus allows its support for legacy protocol stack with minimal variation. Along with the provisioning of QoS, energy management in cognitive radio is also discussed. Since, the goal of the cognitive radio is to enter into a commercial market in a transparent way, energy management of the devices in the secondary network plays a crucial role to achieve this. Energy management from the viewpoint of channel selection, spectrum sensing order, and media access scheme is also discussed in this book. Each of these viewpoints provides different ways through which energy can be efficiently managed in cognitive radio network.

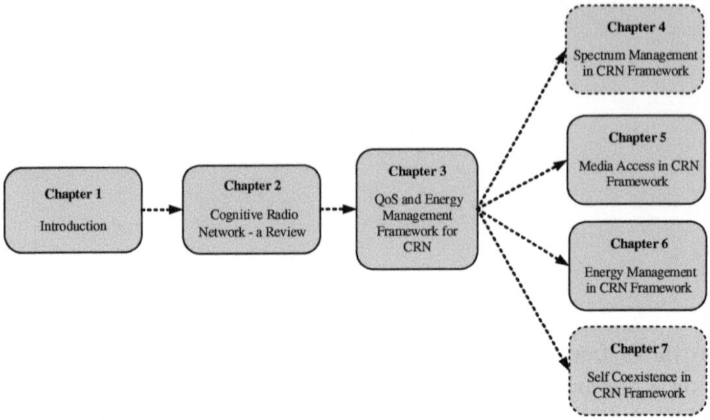

The structure of this book is given as shown in figure above. It is recommended for a beginner to start from Chap. 1, while an experienced reader in the field of cognitive radio can directly proceed from Chap. 3.

Chapter 1 introduces basic motivation for reading this book and provides a basic background on cognitive radio and spectrum regulation issues. This chapter reminds the need of opportunistic spectrum usage and presents a comparison of software-defined radio with cognitive radio. This chapter also introduces basics on QoS, channel selection techniques, MAC protocols, and self-co-existence in cognitive radio network. Chapter 2 provides a review on different modules of the cognitive radio framework, including spectrum management, spectrum selection, media access schemes, and cognitive radio platforms. These modules are discussed from the perspective of quality of service and energy management.

Chapter 3 briefly describes QoS provisioning and energy management framework for cognitive radio network. QoS provisioning framework for cognitive radio network at Layer 2 of the protocol stack and different functional modules for QoS provisioning from the perspective of latency and throughput are discussed in this chapter. It is shown that latency and throughput can be directly identified by the media access scheme and channel selection technique, respectively, and also the effect of self-co-existence and energy management is separately studied.

A case study is conducted on spectrum usage behavior in Chap. 4. The concept of deterministic and stochastic-based classification of channel selection technique is discussed. Deterministic channel selection schemes utilize deterministic behavior of spectrum and make decision on spectrum selection ahead of time. A packet-to-channel tabular mapping-based fast spectrum selection scheme is discussed for stochastic channel environment.

In Chap. 5, a case study is done on a hybrid media access control protocol for QoS provisioning which encompasses two different channel access etiquettes. The discussed media access scheme realizes service differentiation for four different types of packets. The support for important network operations such as broadcast and multicast, as well as realization of power saving mode is also discussed.

A case study is presented on the energy management functionality of cognitive radio network via utilizing a spectrum selection strategy in Chap. 6. A scheme is discussed to allow selection of low bandwidth channels so as to conserve transmission energy. The scheme also realizes a packet admission control technique for graceful degradation of QoS of the system.

Resource relocation and resource sharing-based self-co-existence techniques are discussed in Chap. 7. The approach of graph coloring is utilized to allocate disjoint set of resources, so as to mitigate interference. A frame sharing based approach is also discussed for self-coexistence.

Singapore, Singapore Vishram Mishra
Patna, India Jimson Mathew
Singapore, Singapore Chiew-Tong Lau

Contents

Chapter 1
Introduction

Recently, the rapid proliferation of portable wireless communication devices supporting multimedia content playback and online gaming have emerged in market. The applications supported by these devices consumes hefty bandwidth for which a wider spectrum band is utilized. An example of historical increase in requirement of wider band can be seen in development of market from 2G to 5G generation of mobile networks. The most useful spectrum bands are in the range of sub 3-GHz; which already have been allocated for specific purposes and are regulated by the regulatory authority of the respective country. The conventional method of channel allocation follows *fixed channel allocation* (FCA) policy; which implies allocation of a particular band of spectrum, to an operator/user for a given purpose [1]. The static allocation of channels by FCA policy has resulted in exhaustion of available usable spectrum as the number of bandwidth hungry applications keeps in increasing while there is no room left for available spectrum in the usable band because of overcrowding. This has also resulted in technological advancements to utilize higher frequency bands (60 GHz range); but these suffer from problems of high attenuation and line-of-sight operation.

These large number of portable wireless communication devices utilize spectrum over non-uniform space and time, leading to a significant amount of unused frequency spectrum. Studies sponsored by various spectrum regulators in their respective countries have shown that the more than 70 % of the spectrum allocated to licensed user is not utilized [2]. Therefore there exist a paradox between the underutilization of the spectrum and its scarcity.

The underutilization of spectrum results in spectral holes, also called as white spaces. These spectral holes can be opportunistically exploited by an opportunistic (unlicensed) user, which ultimately leads to a solution for spectrum scarcity. An opportunistic user hops from one spectral hole to another while mitigating interference with incumbent (licensed) users (as shown in Fig. 1.1) [3]. To realize the functionality of an opportunistic user, cognitive radio (CR) based dynamic spectrum access technique has been formulated. The basic idea is to employ a hierarchical

V. Mishra et al., *QoS and Energy Management in Cognitive Radio Network*,
Signals and Communication Technology, DOI 10.1007/978-3-319-45860-1_1

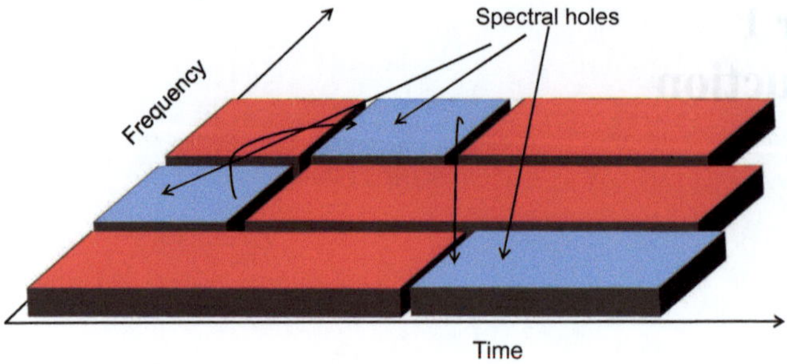

Fig. 1.1 Opportunistic utilization of spectral holes

model, where licensed (primary/incumbents) and unlicensed (secondary/CR) users coexist in the same frequency band such that the primary users are oblivious of the presence of secondary users [4]. A secondary user first senses the channel, and if it is unoccupied then only it performs transmission of data, thereby maintaining interference with primary users to a minimum level.

To accomplish the task of opportunistic utilization, a cognitive radio has to go through the process of *Spectrum Sensing*, *Spectrum Decision* and *Spectrum Sharing* [5]. Spectral holes in the primary bands are detected with the help of spectrum sensing functionality of a cognitive radio. Since results from spectrum sensing are critical in mitigating interference with licensed users, it is of paramount importance to obtain precise information about spectral opportunities. This requires cooperation among different secondary users to avoid false alarm and missed detection [6, 7]. After the detection of spectral holes among different primary heterogeneous bands, a decision is made on allocation of primary channels based on different constraints imposed by the user. Primary bands are governed by local regulatory agencies and hence local policies enforced on them varies [8]. Policies like maximum accessible bandwidth, transmit power and transmit duration are main determining factors for channel selection [9]. The channels are allocated in such a way that the quality of service (QoS) requirements of different applications are met. Different channel allocation strategies try to cater to different needs of users. The performance and competency of a channel selection technique depends upon various factors including number of secondary users, primary user interference, number of channels used and algorithm used [10]. When different users exist on the same channel, an access scheme is required to provide a coordination between different users so that access collisions can be alleviated. Several approaches have been proposed to develop efficient media access control (MAC) schemes for cognitive radio based network [11]. But most of these MAC protocols solve a particular problem and do not provide a generalized framework for cognitive radio system.

Since there is an out-burst of multimedia and voice over IP (VoIP) traffic, an important objective during the operation of network is to ensure a minimum grade of service to such delay sensitive traffic [12]. In a traditional network, all users are designated with equal priority to access a common resource (bandwidth). Therefore if the total demand of resource by traffic from all users in the network exceeds the available resource, then the throughputs of all users are equally impeded. This has varying impact on user experience depending upon the delay-sensitivity of the users application. For example, one second delay is acceptable in web browsing while the same is not acceptable for VoIP users. To mitigate the adverse effects of resource equalization, traffic from different users are prioritized and appropriately handled. Some of the anticipated applications that require special handling of the traffic are VoIP, video streaming, music streaming and interactive gaming [13]. The special handling of traffic is evaluated using quality of service (QoS), which is defined as the ability of a system to provide a guarantee on assured grade of service [14]. A different context to service quality is Quality of Experience (QoE) which is used to describe the perception of the consumer about usefulness of a system [15]. Currently, the development in the field of cognitive radio solves localized problem in their respective domains without any inter-relation. A framework is required which can integrate different modules responsible for QoS provisioning in the cognitive radio system. In this book, case studies are done about a QoS provisioning framework for cognitive radio network which works at the Layer 2 of the protocol stack while integrating the functionality of spectrum decision, medium access control and self-coexistence.[1]

1.1 Why This Book?

The fundamental goal of this book is to discuss a QoS provisioning framework for cognitive radio network which provides QoS support at Layer 2. For the purpose of same, different technologies responsible for QoS have been identified in cognitive radio network and a relationship is established among them. The framework comprises of multiple units, each of which is individually responsible for QoS provisioning in the system as a whole. The coordination among these units (blocks) is discussed to enable the system to perform its intended task. These units are connected in a specified manner, with each connection either uni-directional or bi-directional to indicate the type of relation among different units. The configuration, queries or results are passed from one unit to other unit. The configuration passed to a block contains the parameters to configure it so as to achieve the desired utility from it. The query consists of commands asking for data or result. An example of query would be channel state and its response would be either free or busy. The query-response blocks have bi-directional link between them. The framework works in parallel with the main communication protocol stack and therefore does not directly alter the behavior of system. This results in minimum dependency with easy integration of framework in a conventional communication system protocol stack. The compart-

[1]Layers of the protocol stack refers to the architectural layers of the Open System Interconnection model proposed by the International Organization for Standardization for the purpose of standardization of the internal functions of a communication system [16].

mentalization of different units allows easy abstraction, debugging and maintenance of the system as a whole.

The framework works by providing substitution of protocol stacks in conventional communication system with its own instances. An example would be generation of MAC protocol instance on the fly depending upon the requirement. This will help to classify different activities involved in a QoS framework while dividing the overall work of the QoS framework into smaller units. These units are independently studied, improved and created if they do not exist elsewhere.

The objectives of this book are as follows:

1. Classification and compartmentalization of different modules that are responsible for QoS provisioning at Layer 2 of the protocol stack. The identification of these modules depends upon the type of QoS required: *latency guarantee* or *throughput guarantee*. These guarantees are handled by their respective modules and hence *compartmentalization* allows implementation of abstract behavior of different functionalities provided by the system.
2. Since QoS provisioning modules at Layer 2 are compartmentalized, a relationship need to be defined for the purpose of requirement translation. The guarantee on throughput and latency is translated over these relationship.
3. Each of these modules is independently studied, researched and new techniques are developed to provide better performance than that of its previous predecessor. These modules are incorporated with new interfaces so as to allow interaction with other modules of the Layer 2 of the protocol stack.
4. The performance of each module is independently evaluated which allows clear performance analysis. And it allows developers to give an idea about the maximum performance achievable by the module; so a translation of performance can be done while integrating the module into a system.

1.2 Spectrum Regulation

International Telecommunication Union (ITU) [17] is a specialized agency of the *United Nations* (UN) responsible for the coordination of shared global radio spectrum in the range of 3 KHz–300 GHz as well as the evolution and coordination of different wireless standards. According to ITU each country/state has its sovereign right to regulate its telecommunication. Therefore each country may have its own set of standards laid down besides the standards defined by ITU.

Spectrum regulation may vary from country to country, but they all can be classified into three broad categories:

1. ***Licensed Spectrum***: This type of spectrum is licensed to a particular operator and enforced through a *regulator* [2]. This can be further classified into *exclusive usage* and *shared usage*. Exclusive usage of licensed spectrum allows exclusive usage rights to the licensee, whereas shared usage of licensed spectrum allows two or more operators to utilize licensed channel on a coordinated basis.

2. ***Unlicensed Spectrum***: This type of spectrum is available for free to all users, who conform to the standards defined for accessing the particular band of spectrum. *Unlicensed National Information Infrastructure* (U-NII) radio band is an example of unlicensed spectrum [18]. *Industrial, Scientific and Medical* (ISM) radio band, whose applications including RF heating, microwave heating and medical diathermy machines, is also part of unlicensed spectrum. Recently ISM bands have been widely utilized by non-ISM applications including wireless sensor network, wireless LANs, cordless, *Radio Frequency Identification* (RFID) and *Short Range Devices* (SRD) on various operating frequencies, all for the purpose of communication.
3. ***Open Spectrum***: Open spectrum refers to the methodology of accessing any band of spectrum without any permission under the constrained set of standards, protocols and maximum allowed interference [19, 20]. The overall objective of this technology is to effectively utilize spectrum on spatial basis without causing any harmful interference to operating incumbents.

1.2.1 Licensed Spectrum

Licensed channels are allocated via *Fixed Channel Allocation* (FCA) policy, which implies allocation of a particular block of spectrum to an operator/user for a given purpose [1]. Allocation of channels for exclusive usage results in inefficient spectrum utilization, because even if the spectrum is underutilized or unused, it cannot be accessed by other users. An example of FCA policy is the allocation of 2G and 3G spectrum for cellular communication in various countries. The spectrum allocation chart of Singapore is shown in Fig. 1.2.[2]

Since most part of the spectrum has already been allocated to existing communication technologies, new technologies demand for more bandwidth. Although technological advancements are pushing towards higher frequency bands (in the range of ∼60 GHz), the operation at these bands suffers from high attenuation and requirement of line-of-sight operation. The high attenuation results in requirement of high transmission power making end-user energy inefficient. Similarly, requirement of line-of-sight operation restricts the design and operation of the end-user in a network. To overcome the problem of licensed FCA policy, *spectrum trading* techniques have been proposed as a secondary market for spectrum ownership and usage [21]. In spectrum trading, spectrum can be leased temporarily, sold permanently or traded as Future and Options [22]; aiming to increase spectrum utilization.

1.2.2 Unlicensed Spectrum

Some part of spectrum is allocated by the spectrum regulators for the purpose of unlicensed operation so as to allow users to access spectrum on non-licensee basis. Even

---

[2] Singapore Allocation Chart, Infocomm Media Development Authority. Retrieved from www.imda.gov.sg

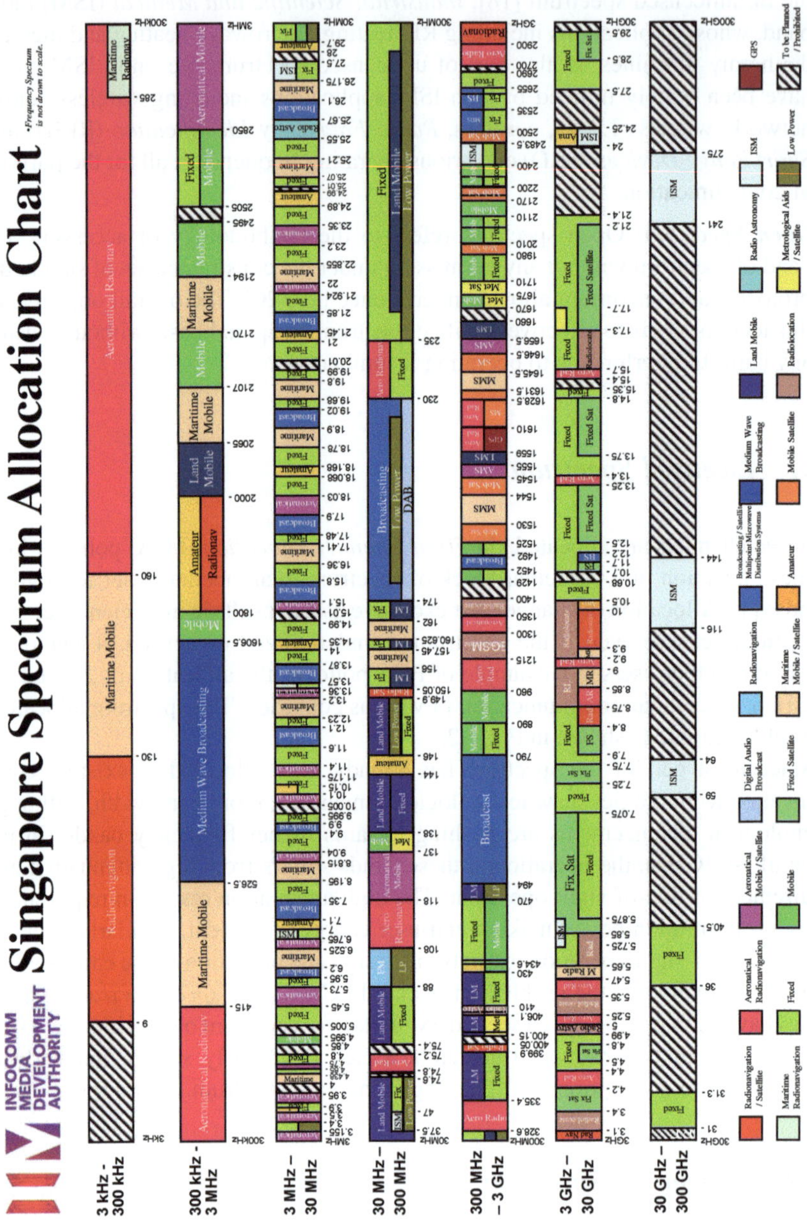

Fig. 1.2 Spectrum allocation chart of Singapore. Almost all frequency bands are allocated up to 30 GHz (Adapted from www.imda.gov.sg)

if this type of spectrum is free-to-use, it is regulated in terms of interference policies which are in turn governed by limitation on transmission power and conforming to the standards of coexistence capabilities. The type of telecommunication users operating on such frequency bands are generally termed as *Short Range Devices* (SRD). SRD generally operate with transmission power less than 100 mW of *Equivalent Isotropic Radiated Power* (EIRP). An example of unlicensed spectrum in Singapore is the allocation of 2.4000–2.4835 GHz band for Wireless Video transmitter, Bluetooth, WLAN and other SRDs [23]. All devices operating in this band are limited to 100 mW EIRP of transmission power, except for WLAN operation where the transmission power is limited to 200 mW EIRP [23].

1.2.3 Open Spectrum

The success of unlicensed bands has resulted in plethora of devices to operate in this band, resulting in overcrowding and hence leading to a problem termed as "tragedy of commons" [24]. In other words, over-utilization of a particular band of spectrum implies high interference thus making it less effective in terms of operation. To mitigate such a scenario, open spectrum techniques are proposed. Open spectrum allows users to access any portion of spectrum under the given constrained set of etiquettes, standards and maximum allowed interference, that need to be enforced for the purpose of spectrum sharing [19]. This allows users to access spectrum without going through the process of licensed spectrum allocation by regulatory agencies. Two major form of open spectrum access techniques have been discussed in the literature:

1. ***Underlay Spectrum Access***: This type of spectrum access refers to the simultaneous secondary uncoordinated utilization of spectrum even in the presence of licensed incumbents on the spectrum. The main idea behind this technique is to keep the transmission power low, so that the interference received by the licensed incumbents is below a certain threshold [25]. To achieve an effective communication, the transmitted signal is spread over a wide band of spectrum with low power transmission. Various spread spectrum techniques can be applied to achieve this including *Ultra-Wide Band* (UWB) which has become popular for its utilization in system such as *Wireless USB* (WUSB) and *Wireless Firewire* [26]. This type of spectrum access do not require development of new protocols.
2. ***Overlay Spectrum Access***: This type of spectrum access refers to the utilization of under-utilized spectrum by a secondary user while mitigating interference with licensed incumbents; a technique referred as opportunistic utilization [27, 28]. To realize the implementation of such secondary user, development of new protocols is required. Cognition capabilities are implemented in the system for the purpose of identification of opportunities and its utilization [29].

1.3 Opportunistic Spectrum Usage

The allocation of spectrum in Singapore is shown in Fig. 1.2, and it might be noted that almost all frequency bands up to 30 GHz have been allocated. Even though all frequency bands have been allocated, they are under-utilized over a given space and time. A recent study conducted by researchers in Singapore has shown that the average utilization of spectrum in 80 MHz–5.85 GHz range is only 4.54 % [30]. The average spectrum occupancy in Singapore is shown in Fig. 1.3.

Spectrum occupancy for different locations all over world have been collected and studied by various researchers. *Shared Spectrum Company* is one that is involved in thorough and wide range of spectrum occupancy studies. The firm has generated many spectrum occupancy reports in the range of 30 MHz–3 GHz that are publicly available [31–39]. A collective study of these reports show that on average at any given place or time, the average occupancy of a particular band of spectrum varies from 0 to 30 %. Various other researchers have shown similar results in the context of average spectrum occupancy across different parts of the world [40–47].

These studies show that there exists a paradox between the scarcity of spectrum and its under-utilization. Under utilization of frequency spectrum implies that white spaces exist in the spectrum; which can be opportunistically utilized by an intelligent opportunistic device [48]. The design of such opportunistic spectrum access follows the principle of overlay spectrum access. Therefore such opportunistic users exist

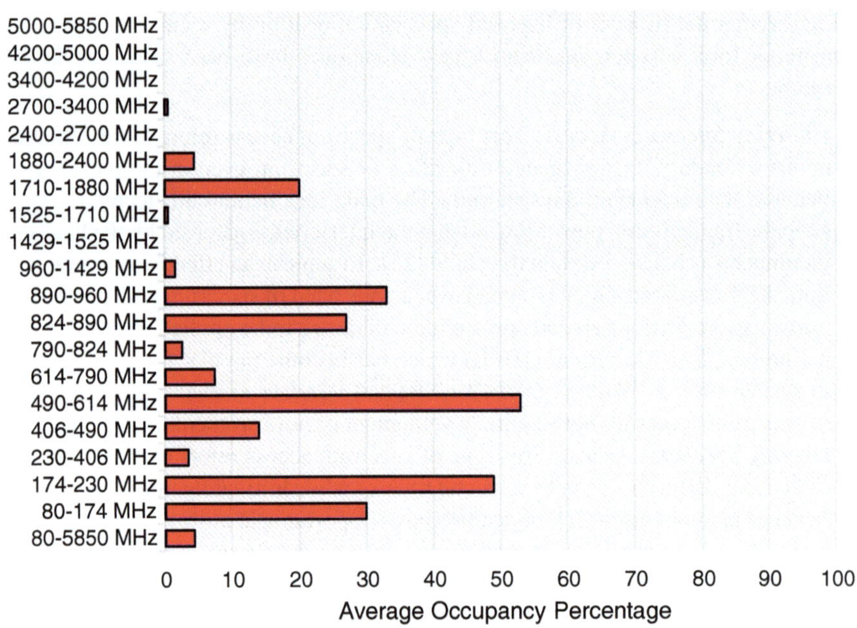

Fig. 1.3 Average spectrum occupancy in Singapore

with licensed users in a hierarchical way, where incumbent users (licensed users) are designated as primary users (PU) while opportunistic users are designated as secondary users [3]. A secondary user (SU) periodically senses a channel for the presence of an incumbent. It performs its own communication only when the channel is free from any incumbent and relinquishes it immediately whenever an incumbent signal is detected on it. Therefore the interference from secondary users are kept minimal such that primary users are oblivious of their presence. The design of such secondary network allows effective utilization of spectrum, but requires cognitive capability to automatically detect spectral holes in the spectrum and utilize them on transparent basis [29]. This requirement is the key concept of *Cognitive Radio*.

A recent trend in wireless communication is towards realization of fully programmable wireless communication system through computer software. This allows cellular telephony network operators to cope with the ever changing standards of cellular technology. Various professional and research organization have shown interest in *Software Defined Radio* (SDR) that not only solves the problem of coexistence of heterogeneous networks, but also alleviates the problem of inefficient spectrum utilization. The ongoing projects *Wireless World Initiative New Radio* (WINNER) [49, 50] by European Union and *NeXt Generation Communication Program* (XG) [51, 52] by *Defense Advanced Research Project Agency* (DARPA) of United States are focussing on the development of technologies and frameworks for flexible radio-interface and network architecture so as to improve the spectrum utilization. The committee *Dynamic Spectrum Access Network* (DySPAN) [53] is one of the standards committee sponsored by IEEE, aimed at developing standards for radio and spectrum management. Several working groups of DySPAN have been formulated to tackle multiple objectives of efficient spectrum utilization.

1.4 Software Defined Radio and Cognitive Radio

Conventional radio communication systems have components implemented in hardware, which restricts the flexibility of the system while reducing the cost of deployment. A recent change is seen towards the realization of wireless communication system completely in software. This allows cellular operators to deploy new communication technologies via a software update rather than changing the whole/part-of radio platform. A *Software Defined Radio* (SDR) is realization of all transceiver capability as a software program on an appropriate computing platform (processor) [54]. As an evolutionary step, a SDR may implement band filters on hardware. Most of the work done on SDR is from the perspective of development of Physical layer of the Open Systems Interconnection (OSI) protocol stack. But with the requirement of cross-layer optimization for the purpose of effective communication system, SDR has evolved with the development of complete protocol stack.

An opportunistic utilization of the spectrum for the purpose of dynamic spectrum access will require reconfigurability in waveform and protocol. Therefore an SDR along with appropriate cognition capabilities can be recognized as a potential candidate for secondary network operation. The cognitive concept for dynamic spectrum access (DSA) was first introduced by Joseph Mitola during his work on SDR [27, 28, 48]. By definition, a "Cognitive Radio" is a radio that can change its transmitter parameters based on interaction with the environment on which it operates [3, 5]. A cognitive radio has the capability to develop waveform and generate protocols in real-time using various cognitive capabilities. Basically, it is an enhancement to SDR, which builds up the reconfigurable radio communication system. FCC has identified certain features that can be implemented via cognitive radio to effectively utilize the spectrum [3]:

1. **Frequency Agile**: The radio utilized by cognitive radio is able to change its operating frequency so as to effectively utilize the spectrum while mitigating potential interference to the licensed users.
2. **Reconfigurable Modulation**: The transmission parameters and generated waveform can be reconfigured to adapt to a certain operating environment. This is required for selecting a robust modulation scheme for the designated environment which also conforms to the standards defined by regulatory authorities for shared spectrum usage.
3. **Transmission Power Control**: The radio can dynamically adjust its transmission power from maximum allowed EIRP to low power transmission. The operation at maximum allowed EIRP allows effective and robust communication, while operating at low transmission power allows improved sharing of the spectrum.
4. **Geolocation Awareness**: The radio is enabled to identify its location along with the knowledge of geographical distribution of spectral allocation. This assists in spectrum sensing to identify spectral opportunities in an effective way. Cooperative geolocation awareness of other cognitive radio users allow a radio to effectively manage the shared spectrum.
5. **Mutual Coordination**: A coordination based channel sharing can be achieved between a licensed incumbent and a secondary user. This coordination can be pre-defined or it can be spawned on a real-time basis.

Since a cognition enabled software defined radio follows open spectrum sharing policy, it can share spectrum with *licensed* spectrum or *unlicensed* spectrum users. The former type of sharing is called as *vertical* sharing while latter as *horizontal* sharing [55]. Therefore vertical sharing refers to the overlay spectrum sharing of licensed spectrum with primary incumbents. Similarly, horizontal sharing refers to the overlay spectrum sharing of unlicensed channels with *Short Range Devices* (SRDs). A horizontal sharing may also refer to the sharing of unlicensed spectrum by two or more heterogeneous cognitive radio user.

1.4.1 IEEE Groups Working on Spectrum Sharing

Over the years, several IEEE working groups have been formulated for the purpose of effective spectrum sharing. These groups are designated to increase the utilization of spectrum for the mentioned communication technologies.

1. **802.11a**: This group is responsible for designing *Dynamic Frequency Selection* (DFS) scheme so as to allow unlicensed users (IEEE 802.11a) to share spectrum on an overlay basis with primary incumbents such as radar stations [56]. According to the standards defined, an 802.11a device while operating on a U-NII band, should perform periodic scan of channels to mitigate interference with radar systems.
2. **802.11k**: This group is intended to provide standards for measuring spectral opportunity while providing means for measurement, reporting, estimation and characterization of spectrum usage [57]. The measurement values can be utilized by an opportunistic user.
3. **802.16.2**: This group is responsible for mitigating interference and providing coexistence capabilities to *Wireless Metropolitan Area Network* (WMAN) systems based on IEEE 802.16 standards which operate on licensed bands [58].
4. **802.16h**: Similar to IEEE 802.16.2, this group is responsible for mitigating interference and providing coexistence capabilities to *license-exempt* systems based on IEEE 802.16 standards [58]. The standard follows the approach of hierarchical overlay sharing of spectrum by a secondary user.
5. **802.19**: This group is known as *Coexistence Technical Advisory Group*, which is responsible for designing policies for coexistence between different standards of communication [58]. Therefore to evaluate the coexistence capability of any new upcoming radio communication standard, the group can provide analysis to check its conformance to the policies defined in IEEE 802.19 standards.
6. **802.22**: This group is responsible for standardization of wireless broadband internet services via vertical spectrum sharing with TV bands in the range of 54 MHz–862 MHz [59]. This service is given the name of *Wireless Regional Area Network* (WRAN), and it utilizes overlay spectrum sharing with white spaces in TV bands. Since operation of cable and fiber services for broadband access is not economically viable in rural/sparsely-populated areas, WRAN is developed as an alternative.

1.4.2 Cognition Cycle

Since a cognitive radio is based on its capability to sense the external environment and act accordingly, a form of cognition is required for its realization. The cognition cycle comprises of various states through which a cognitive radio cycles to continuously observe the environment and decide on the best course of action [5]. The observation of environment indicates detection of primary incumbents on different spectrum, while actions indicates making a decision on selection of a particular spectrum band for transmission [28]. This decision also include selecting modulation scheme, transmission frequency, power, rate etc. An analysis is done before making a spectrum decision, which comprises of different environment and network parameters, including channel noise, transmission power, number of primary and secondary users, spectral bandwidth, bandwidth required, SNR required, etc. A cognition cycle is shown in Fig. 1.4.

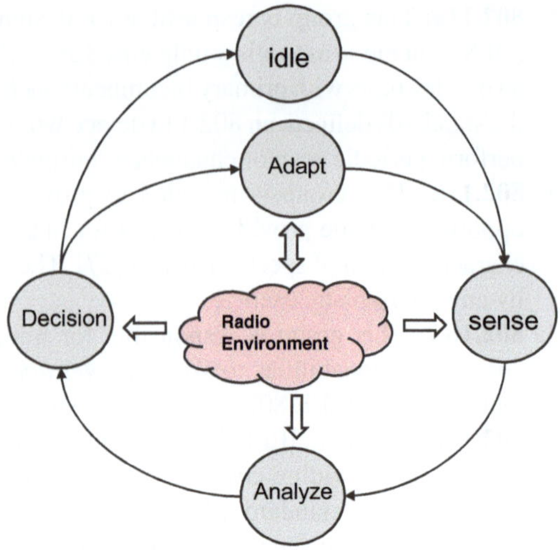

Fig. 1.4 Cognition cycle for cognitive radio

The cognitive radio starts from the idle state and goes into environmental sensing state. During this state, it records all the environment parameters it senses. It analyzes these environment parameters in the next phase to get more logical and physical figures. Based on the obtained radio environment parameter and the required operating conditions (required SNR, bandwidth etc.), the cognitive radio makes a decision in the next state. The decision is to either adapt the cognitive radio system to the desired spectrum band with operating behavior and characteristics (center frequency, bandwidth, tx power, rx gain etc.) or to go into idle state. The adaptation state is applicable for both, cognitive radio operating in transmitting or receiving mode. If no suitable spectrum is found, then the cognitive radio goes into idle mode (sleep mode) and waits for next sensing cycle.

1.4.3 Cognitive Engine and Framework

A cognitive engine (CE) is formulated and implemented on various SDR boards in [29] while utilizing following components:

1. **Perception**: Sensing done by sensors collect the internal and external data. Internal data consists of resources available like waveform capabilities, computational power, battery level etc. External data consists of channel availability, radio regulations and user demands.
2. **Comprehension**: Part of the system that comprehends and combines the knowledge along with the sensed data so as to aid adaptation mechanism.
3. **Execution**: Optimization and adaptation mechanism to control the radio behavior.

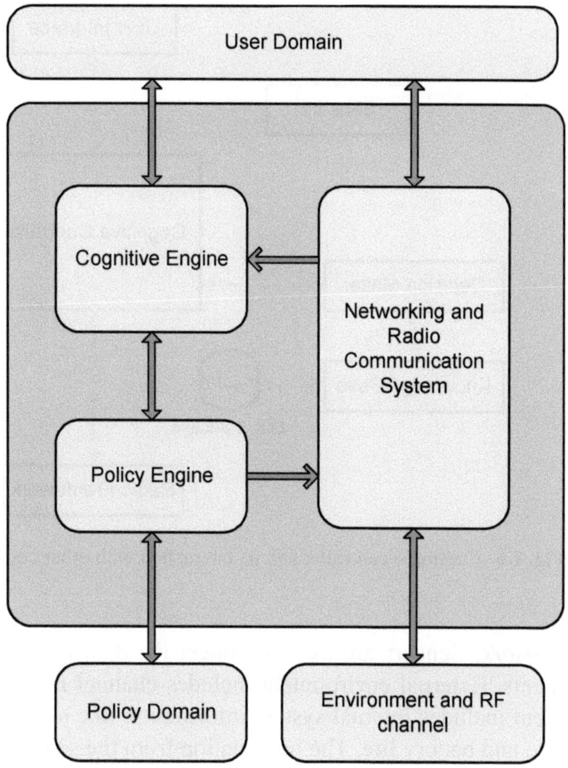

Fig. 1.5 A typical cognitive engine

A cognitive engine, as shown in Fig. 1.5, is a separate entity that works alongside the normal communication path. It only tells how to best control the communication path. User domain conveys the information about the performance requirements of services and applications (QoS) running on top of the system to the CE. External environment and RF channel provide the environmental context to the radio transmission and reception behavior. For example a multipath environment requires a complex receiver than a single line-of-sight propagation model. The external environment information also helps in providing optimization boundaries on the decision-making and waveform development. Policy domain restricts the system to work within the working boundaries set by the regulatory authorities like FCC. These policies might restrict the usage of particular spectrum, the maximum amount of power of a radio etc.

A typical architecture of cognitive engine developed by Virginia Tech's Centre for Wireless Telecommunication is shown in Fig. 1.6 [29]. It comprises of a cognitive controller attached to components like sensors, optimizer, decision maker, policy engine, radio framework and user interface. The Cognitive controller acts as the system kernel for the architecture and schedules the input/output and timing of the other attached components which are discussed below:

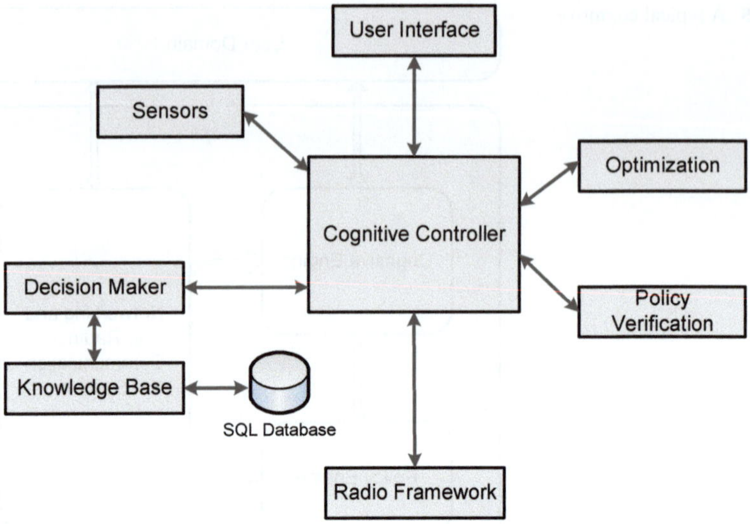

Fig. 1.6 Cognitive controller and its interaction with other components

Sensors: Sensors are used to collect the data from external and internal environments. External environment includes channel noise/availability. Internal environment includes internal system information like noise power, signal power, bit error rate and battery life. The information from the sensor component can be used by an optimization process or a decision based process.

Optimizer: The optimization process is used to maximize the performance of the system by selecting the most appropriate mode of communication. The parameters required in the optimization process are user QoS requirement, physical environment, computational resources available and radio regulations. Given a required QoS, the optimizer selects an appropriate method of communication which satisfies with the other parameters.

Decision Maker: This component receives the information from the sensors and makes decision on the basis of the information. It also decides when a reconfiguration is needed and whether an optimizer is required, and if required what are its goals. It can also use past information to help the optimization process. In general, case based decision theory is used for decision making purposes.

Policy Engine: The waveform selected/created by the optimization process is checked to see whether it is legal to be used in that location over the given time. The policy engine consists of a database of the regulatory policies. Whenever a waveform is selected/generated, it is checked against the database that it is within the bound. The policy engine is dynamic and evolves with the time to accommodate any change done by the regulatory authorities.

Radio Framework: Radio framework acts as a middleware in between cognitive engine and the radio platform. It translates the generic information into the implementation specific information. All the information such as frequency, modulation type, bit rate, packet size, are passed to the radio platform using a simple generic representation with the help of a radio framework.

User Interface: The user interface acts as a control center for the whole cognitive engine. From this interface a user can control the engine as well as provide parameters for the working of the system.

Cognitive Controller: Cognitive controller acts as a scheduler and a system kernel to schedule and handle various components attached to it. It utilizes different implementation capabilities of the components attached to it.

A work on the development of a framework for cognitive radio from the perspective of media access control is reported in [60]. The authors developed a MAC framework for providing composition of different MAC protocols best suited for a given application, available radio and given spectrum regulatory policies. MAC protocols are decomposed into their basic functional components and are wired on the fly to realize envisioned MAC protocol. MAC functionalities are implemented as a library so that different types of MAC protocols can be envisioned by connecting different modules of the library in an appropriate manner. A wiring engine is used to bind the blocks together and hence provide control and data information among blocks. Another work on MAC framework development is done in [61] where authors proposed an Intermittently Connected Cognitive Radio MAC (ICR-MAC) framework. The framework provides support for cluster formation of the intermittently connected network and tries to maximize the transmission opportunities. The ICR-MAC framework supports multiple flow-scheduling, i.e. within a cluster more than one channel can be utilized for communication purposes.

1.4.4 Cognitive Radio Network

The network formation by cognitive radio system can be classified into two types: *Infrastructure based* CR network and *Cognitive radio based ad-hoc network* (CRAHN) [5, 59]. An infrastructure based CR network has a central network entity such as an access-point (AP) or a base station (BS), whereas CRAHN does not require any such central network entity, and therefore users in CRAHN communicate with each other on *ad-hoc* basis.

Users in an infrastructure based CR network follow the rules and etiquette set by the central network entity which also acts as a controller. The users of such network observe and analyze the spectrum behaviour and transmit this information to the base-station, so that it can make an appropriate decision. Based on the decision of BS, users may have to reconfigure their communication parameters so as to remain effective in the CR network. Figure 1.7a shows 4 network elements along with an access point. The access point may also act as a gateway for external world communication (internet).

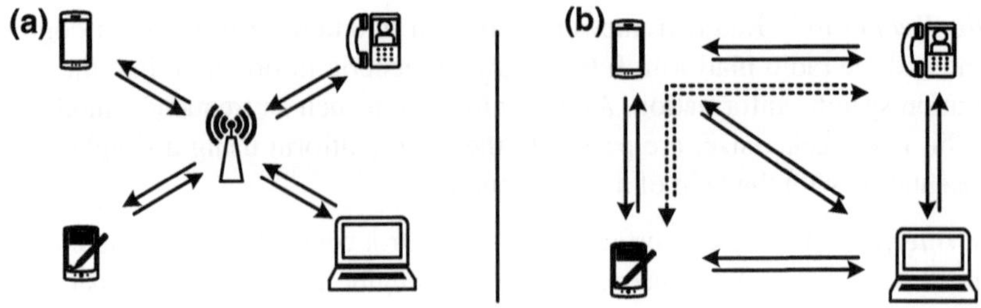

Fig. 1.7 Different types of CR networks: **a** Infrastructure based CR Network. **b** Cognitive Radio based ad-hoc network (CRAHN)

Users in CRAHN follow the standards and etiquettes defined for the formation of the *ad-hoc* network. Therefore all users in such a network cooperate with each other to achieve a common objective. Users in such network show collaborative behavior and exchange the information about the spectrum behaviour which they learn through observation. If there is any reconfigurability required, all users in the network have to agree on it. This can be done either by having a voting system, or through a predefined etiquette which is followed by all users in the network. Users in such a network, if within communication range, directly communicate with each other. If users are not within communication range, then the message passing is done through *multi-hopping*. In multi-hopping, the message traverses through the sender to destination with the help of intermediate users (part of routing scheme). Figure 1.7b shows a CRAHN operation with 4 users in the network. The dotted line in the figure indicates the routing required in the absence of a direct communication path or when a direct communication path is not economically suitable.

1.5 Quality of Service (QoS)

In a traditional wireless network, all devices are given equal access and priority. If the total demand of bandwidth resource from the traffic generated by all users in network exceeds the available bandwidth, then the throughput of traffic from all users is equally impeded [14, 62, 63]. This has varying impact on user experience depending upon the type of application. For instance, a second delay of page loading in web browsing application will not affect user experience; while the same delay experienced by a VoIP user will result in inconvenience. Therefore to avoid inconvenience to the user, a certain guarantee of resource is provided to the application. The guarantee can be provided in terms of throughput, latency, packet drop rate, and service response time. This is defined by *Quality of Service* (QoS), which defines the ability of a system to guarantee an assured grade of service. To achieve a defined QoS in a system with static resource, packets are prioritized in order of their user experience requirements. For instance, a VoIP packet will have high priority over an HTTP (web page) packet. A higher priority packet gets higher chance/precedence of packet transmission over low priority packets. Therefore when the total demand of bandwidth resource from the traffic generated by the users in network exceeds the available bandwidth, then a higher priority packet is given the precedence for transmission, while blocking low priority packets. This results in effective bandwidth utilization as a result of low contention and hence low packet collision [13]. Also a high priority packet gets the transmission opportunity thereby maintaining the required QoS. All of the above is realized with the help of blocking of low priority packets, which can tolerate certain amount of delay. The blocked packets are released when the transmission of higher priority packet is completed.

A related term in the context of QoS is *Quality of Experience* (QoE). QoE is a subjective term and is used to describe the experience/satisfaction of a user with a particular service/system [15]. QoE and QoS are mutually dependent on each other and in order to achieve a particular level of QoE, some bound on QoS must be guaranteed. A typical subset of QoS in telecommunication is defined as Grade of Service (GoS). GoS is used to define the blocking/dropping probability of the calls made in a system. Generally GoS is calculated by taking the ratio of total number of calls dropped or blocked to the total number of call attempts made.

QoS provisioning is of crucial importance for network supporting real-time traffic. For instance, in a health-care monitoring system, the results from various sensors are aggregated and transmitted on a real-time basis. This allows servicing any emergency/urgency required by a patient immediately at the time of action. If real-time traffic is multiplexed with non-real time traffic, packet prioritization or QoS provisioning is required [64]. To implement QoS provisioning following techniques need to be implemented by a system:

1. **Access Category**: To provide packet priority, packets need to be classified into different categories. These categories are termed as *Access Category* (AC). An access category refers to the priority *category* of a packet to *access* a resource. The WMM profile in IEEE 802.11e defines 4 types of access category [14].
2. **Traffic Classifier**: To identify the access category (AC) of a packet, some kind of identification scheme is required. In general, packet marking is required to identify a particular type of data packet, which is in general done by application. But if an application does not support packet marking scheme, then a traffic classifier is required to classify a packet [65].
3. **Packet Admission Control**: Once the packets are identified, a high priority packet is allowed to access the resource while a low priority packet is blocked. This is achieved via packet admission control technique, which follows certain rule to guide packet admission, while also mitigating the problem of low priority packet starvation [66].
4. **Starvation Prevention**: If low priority packets in a system are always blocked due to excessive high priority packets in the network, this results in resource starvation of low priority packets [67]. Therefore the user experience for low priority packets may degrade, resulting in low QoS of the system. Therefore a scheme is required to prevent starvation of low priority packet while maintaining a graceful degradation of QoS.

5. **Resource Requirement Identification**: A packet in general may require a high priority in the context of either *latency* or *throughput*. This is discussed in Chap. 3, Sect. 3.1.1, where resource requirement identification of a packet is done on the basis of its characteristics. If a packet requires high priority in the context of latency, it implies that it gets the first access to the available resource than its counterpart packets [68]. On the other hand, if a packet requires high priority in the context of throughput, it implies that the packet is transmitted over a non-shareable resource which has enough bandwidth to satisfy its throughput requirement [69]. If the packet is transmitted over a shareable resource, a reservation (guarantee) is done to make sure the packet receive enough bandwidth to satisfy its throughput requirement.

As discussed above, resource identification is done to determine the kind of resource requirement needed by a packet so as to maintain acceptable user experience. The following sections discuss on how to implement QoS provisioning in the context of different resources.

1.5.1 QoS Provisioning for Latency Guarantee

Traditional IPv4 network supports a level of prioritization to traffic using Type of Service (ToS) field in IP header which is replaced by *Differentiated Service Code Point* (DSCP) field. This implementation is termed as *Differentiated Services* (Diff-Serv) [12, 13]; which provides low-latency services to time-critical (real-time) traffic while coexisting with best effort traffic. This is implemented at the Layer-3 of the protocol stack and hence is able to define packet-forwarding (routing) behavior via various DiffServ classes. DiffServ defines the *Assured Behavior* with four separate classes, each having three categories for drop precedence (Low/Med/High) of packets. The categorization into separate classes and drop precedence allows the DiffServ to control the routing and retransmission delay of individual packets. Therefore in a congested network, a router enabled with appropriate DiffServ will drop a low priority packet as defined in the policies, so as to allow a high priority packet to access the resource. This mechanism also allows selection of queuing algorithms like: *Priority Queuing*, *Fair Queuing* and *Weighted Fair Queuing* (WFQ) [70] allowing a network operator to control the behavior of queuing delay of packets.

In wireless networks, IEEE 802.11e working group is responsible for the provisioning of QoS in IEEE 802.11 networks [71]. It defines enhancement for the prioritization of packets that are targeted at MAC layer of wireless network. The basis of QoS prioritization is the introduction of new coordination function: *Hybrid Coordination Function* (HCF). HCF utilizes two method of channel access: *HCF Controlled Channel Access* (HCCA) and *Enhanced Distributed Channel Access* (EDCA) [72]. HCCA defines the central coordination based contention-free access of channels, while EDCA defines the distributed contention based access of media. Therefore HCF has the capacity to divide a MAC superframe into two separate durations:

Table 1.1 Contention window size of different access category [72]

AC	Type	CW_{min}	CW_{max}
AC_VO	Voice	$\frac{aCW_{min}+1}{4} - 1$	$\frac{aCW_{min}+1}{2} - 1$
AC_VI	Video	$\frac{aCW_{min}+1}{2} - 1$	aCW_{min}
AC_BE	Best effort	aCW_{min}	aCW_{max}
AC_BK	Background	aCW_{min}	aCW_{max}

Table 1.2 Typical values of EDCA parameters for $aCW_{min} = 15$ and $aCW_{max} = 1023$

AC	AIFSN	Default CW_{min}	Default CW_{max}
AC_VO	2	3	7
AC_VI	2	7	15
AC_BE	3	15	1023
AC_BK	7	15	1023

Contention free period to be utilized by HCCA and *Contention period* to be utilized by EDCA. IEEE 802.11e defines four access categories for packet prioritization. The packets are prioritized by defining different arbitration inter-frame spaces and contention window size. An *arbitration inter-frame space* (AIFS) refers to the time duration which a user has to wait before transmitting a packet. A shorter AIFS gets higher probability of transmission and vice-versa. The value of AIFS is obtained with the help of *Arbitration Inter-frame Space Number* (AIFSN) and is calculated as [72]:

$$AIFS[AC] = AIFSN[AC] * aSlotTime + SIFS \tag{1.1}$$

Here *aSlotTime* is the size of time-slot of the system time, while SIFS is the size of short inter-frame space defined for highest priority packets like acknowledgement (ACK). Table 1.1 shows the maximum (CW_{max}) and minimum (CW_{min}) contention window size of different access category of packets defined for a typical aCW_{max} (maximum) and aCW_{min} (minimum) values utilized by the system respectively.

For a typical value of $aCW_{min} = 15$ and $aCW_{max} = 1023$ in 802.11e PHY, Table 1.2 shows the default contention window parameters of EDCA.

A shorter value of AIFS allows a high priority packet to win the arbitration period and gain the opportunity for transmission. Apart from contention based data transfer, IEEE 802.11e also provides contention-free access of channel by utilizing a period called as *Transmission Opportunity* (TXOP). A TXOP is a maximum bounded time interval during which a user may send as many packets as it intend to. The TXOP is only defined for voice (AC_VO) and video (AC_VI) data packets.

Wi-Fi Multimedia (WMM) [14], is an alliance for certifying interoperability of devices based on IEEE 802.11e standard. It defines the standard of access categories for packet prioritization, however it lacks the standard for providing guaranteed throughput. Therefore throughput guarantee is a separate issue that needs to be taken care of.

1.5.2 *QoS Provisioning for Throughput Guarantee*

Throughput guarantee is separately handled by a system providing QoS. Throughput may indirectly affect the latency of a traffic from a user [73]. For instance, if the throughput requirement of a traffic exceeds the available bandwidth resource, then the packets need to queue which increases the latency of the system. In a video-on-demand system, if the communication media is insufficient to support the data traffic, then the video playback suffers from frequent buffering. In a packet-switching network, a guarantee on assured level of throughput can be provided via: *Resource Allocation* and *Packet Admission Control* [66, 74].

1.5.2.1 Resource Allocation

Resource allocation refers to the allocation of wireless resources to all the users in a system [75]. A resource allocation scheme may provide a guarantee on what kind of resource is allocated to a particular user. A user with high demand for throughput may be allocated a high bandwidth resource. For instance, a user who frequently watches high-definition content on a wireless device may be allocated a high bandwidth channel to support the traffic throughput. In a multi-channel environment, resource allocation translates to allocation of channels to users. Therefore a user may be allocated with one or more than one channel to accommodate his need. There can be multiple objectives of a resource allocation scheme including maximizing the system throughput or providing fairness among users in the network.

1.5.2.2 Packet Admission Control

In a resource constrained environment where resources cannot be allocated or distributed, for instance a network operating on a single channel, the available resource is reserved by having packet admission control scheme. A packet admission control scheme allows a higher priority packet to access the resource, while a low priority packet gets blocked [76]. Therefore this ensures that the resource is entirely utilized by high priority packets and hence is guaranteed to be available.

1.6 Channel Selection Techniques in Cognitive Radio Network

The performance of a network depends upon the bandwidth and quality (SINR) of the wireless spectrum (channel) onto which it operates. For cognitive operation of a system on open spectrum sharing basis where channel availability sporadically varies, selection of channels play an important role in determining the quality of service of network operation [77]. For instance, selecting a channel whose availability varies frequently will cause network to switch channels frequently resulting in jitter and increase in packet loss. Therefore channel selection is an important task that need to be performed by spectrum manager of the cognitive radio system while considering various parameters like channel availability, SINR, bandwidth, path loss and primary user behavior.

Spectrum decision is a three step process: *spectrum characterization*, *spectrum selection* and *system reconfiguration* [78]. Spectrum characterization involves identifying different characteristics of spectrum from local and statistical observation including: *path loss*, *error rate*, *interference*, *switching delay* (hardware dependent) and *capacity*. Based on regulatory policies additional characteristics like *maximum holding time*, *allowed bandwidth* and *allowed maximum EIRP* need to determined. Spectrum selection involves selection of appropriate spectrum based on the characteristic obtained and the demand of QoS of secondary users. System reconfiguration involves adapting transmission parameters of the transceiver so as to operate within the selected band of spectrum.

There have been several attempts in developing efficient channel selection algorithms which provide a channel or a set of channels satisfying certain constraints [3, 78]. These constraints are normally applied on the characteristics of the given spectrum. Spectrum decision can be classified on the basis of *network mode* of operation. In *infrastructure network* mode of operation, channel selection decision is made by the central network entity: base-station or access point. Based on the network etiquettes, single or multiple channels can be selected by the base-station and allocated to different end-users in the network. In *ad-hoc network* mode of operation, channel selection decision is made by individual users on collaborative behavior basis. Each communicating pair in *ad-hoc* network selects a channel which is orthogonal to the channels selected by other communicating pairs. This allows multiple pairs to simultaneously perform communication without interference.

1.6.1 Channel Selection in CR Based Infrastructure Network

As discussed in Sect. 1.4.1, IEEE 802.22 WRAN standard has been formulated for the purpose of broadband provisioning in rural area via vertical spectrum sharing on TV bands with the help of cognitive system. In [8], dynamic channel management for IEEE 802.22 has been discussed from the perspective of protection of incumbents from secondary network activity. The work describes the operation of *channel sensing*, *channel selection* and *channel information maintenance* for dynamic channel management. Since WRAN operates on TV white space which might operate with a bandwidth of 6 or 8 MHz (depending on region), the system is able to adapt its bandwidth. A channel width adaptation algorithm is presented in [79], which can maximize throughput or minimize the power consumption. A detailed analysis is done to show the impact of channel width on throughput, range and power consumption. An adaptive spectrum assignment algorithm is discussed in [80], which handles the spectrum variation and fragmentation. The technique is called signal interpretation before Fourier transform and it reduces the time to detect transmission in variable spectral width system by analyzing raw signals in time-domain. In [81], a game theoretic framework has been discussed, which models the channel selection as a non-cooperative game. Spectrum bandwidth, interference and spectrum opportunity holding time is utilized for the purpose of defining the cost function. Cost minimization is done and the spectrum with the minimum cost is selected as the best spectrum for the secondary user. A pre-defined policy based channel selection framework is discussed in [9]. The framework utilizes multiple criteria including minimum SINR, average SINR, coverage area, occupancy history, number of supportable nodes and incumbent appearance probability to select multiple channels. Three different schemes have been discussed for channel selection depending upon the weights assigned to the criteria, selection order and a combination of these. In [82], channel switching delay is taken into consideration for channel selection. Various techniques including geolocation historical information is utilized to avoid selecting previously occupied channels. A trade off between higher throughput (bandwidth) channels and switching overhead is studied in [10]. The results show that utilizing higher throughput channels (with large bandwidth) with few secondary users increases the switching overhead due to increase in sensing time.

1.6.2 Channel Selection in CR Based Ad-hoc Network

A centralized and distributed protocol for spectrum allocation is discussed in [83]. The concept of time-spectrum block is introduced to model the spectrum reservation by a secondary user. Here a portion of spectrum is reserved for a particular time to a user over a given region, such that the communication does not interfere with the incumbents. In [64], channel selection based on switching techniques is discussed. Two channel switching techniques are proposed in case the current channel becomes unavailable: *periodic* switching and *triggered* switching. In periodic switching, a CR user switches to a new channel at the start of pre-defined interval, while in triggered switching a CR user immediately switches to a new channel whenever the current channel becomes unavailable. In [84], the authors have proposed a game theoretic based channel selection scheme for cognitive radio based *ad-hoc* networks. Two different objective functions of game have been proposed to capture the utility of selfish and cooperative users. Utility function for *selfish* user accounts for interference on a given channel, while utility function for *cooperative* user accounts for interference on a given channel and the interference it will cause to other users. A potential game has been formulated with the help of objective functions and is shown to converge at Nash equilibrium point. A learning automaton based channel selection algorithm is proposed in [85]. The algorithm optimally selects a channel which maximizes the probability of successful transmission and reduces the number of costly channel switching. In [86], the authors propose a metric to characterize the channel quality and use it to select the channel. The Busy Time Ratio (BTR) metric uses collaborative sensing and thus reduces the sensing overhead by allowing each CR user to scan only the channel in which they were already performing some communication. In [87], an automatic channel selection algorithm has been proposed which co-exists with a Wi-Fi network. The algorithm selects a channel based on measured utilization of the ISM band while also considering the QoS requirement of the CR users. The algorithm can also achieve user selectable trade-off between system throughput and users' fairness. In [88], the author compares the multi armed bandit (MAB) problem with that of the channel selection in cognitive radio. The author reviews and performs comparisons of algorithms for finite time MAB in static and dynamic environment. A QoS based framework is discussed in [89] which utilizes the bandwidth delay product as a measure for selecting the optimal channel so as to meet the user's QoS requirement. In [90], the authors discussed a distributed channel selection protocol for single hop transmission using two transceivers and a common control channel. The protocol facilitates load balancing to the system and thus decreases cluttering on the same channel. Graph coloring based resource allocation techniques can be utilized for allocating channels among users of CR based *ad-hoc* network. The graph coloring based resource allocation scheme are discussed from the perspective of self-coexistence in Sect. 1.8.

In case of multiple cognitive radio network operation in a given region, the channel selection scheme needs to mitigate the *channel selection collision*. Channel selection collision implies selection of the same channel by two or more different networks which may result in *access collision*. This is discussed in more detail in Sect. 1.8.

1.7 MAC Protocols for Cognitive Radio Networks

After a channel or a set of channels has been selected by a user in cognitive radio network, users perform communication on the selected channel. If two or more users in the vicinity utilize the same channel for communication, it results in *access collision* as shown in Fig. 1.8 [91]. This can be mitigated with the help of an access etiquette which needs to be followed by all users in the network so that frequency-temporal usage among all user pairs are orthogonal to each other in a given region. The technique is termed as *media access control* (MAC). This can be achieved via random access schemes or time-slotted schemes. The MAC scheme utilized by cognitive radio network differs from traditional wireless network in the sense that it incorporates two major design factors: *quiet-period* and *multi-channel* environment [11]. Multi-channel environment allows the operation of MAC scheme to switch to another channel instantly whenever a primary incumbent returns back to the current channel. Quiet-periods are utilized for the purpose of spectrum sensing and are discussed in detail in Sect. 3.4.2.

In a cognitive radio environment, users can choose any of the available channels and hence sender and receiver need to agree upon a common communicating channel which is available to both. For this purpose, initial handshake signals are required to negotiate the communication channel. These signals are termed as control signals,

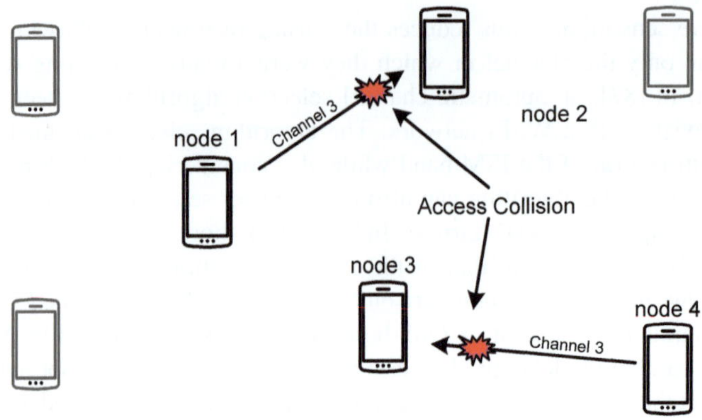

Fig. 1.8 A network of nodes operating in a given region. Nodes 1 and 4 are trying to communicate with nodes 2 and 3 on the same channel, which results in access collision

and since control signals are exchanged between sender and receiver, the design constraint requires them to be on the same signalling channel [92]. This channel is termed as common control channel, which additionally increases the resource requirement of the network by having an additional channel for the purpose of control signalling. This is termed as common control channel problem. Therefore the resource complexity of a MAC scheme depends upon whether it utilizes a control channel or not. An additional resource complexity can be defined in terms of hardware requirement of the MAC scheme: i.e. number of transceivers required. Some MAC schemes utilize single transceiver while others require the usage of multiple transceiver [11]. The MAC scheme differs for different network *mode* of operation and hence *infrastructure* based CR network and *CRAHN* will employ different MAC schemes.

1.7.1 Random Access Based MAC Scheme

Random access schemes do not require time synchronization and are generally based on the principles of carrier sense multiple access with collision avoidance (CSMA/CA). In [91], a random access protocol based on CSMA/CA is discussed for infrastructure based CR networks. The discussed scheme utilizes one single transceiver and closely integrates with the PHY layer to control the transmission power and the throughput of the CR network. Therefore the MAC scheme also supports some form of coexistence with the primary system following classical CSMA scheme. Multiple random access schemes have been discussed for *ad-hoc* cognitive networks. In [93], a distributed MAC scheme has been discussed for users with hardware constraints of single transceiver and limited spectrum sensing capability. The protocol performs limited *fine* spectrum sensing and makes a decision on the number of channels that need to be scanned. The protocol utilizes a common control channel for the purpose of signalling by performing contention on it. The winning pair of contention then performs channel sensing on the given channels. Based on stopping rule for spectrum sensing, the pair may perform transmission or start sensing a new channel. A single transceiver based MAC scheme is also proposed in [94] which utilizes a method of cross-channel communication, in which the sender and receiver utilize different spectrum for sending and receiving. Therefore large spectrum block can be allocated for sending purposes while small spectrum block can be allocated for acknowledgement purposes. An opportunistic spectrum access (OSA-MAC) scheme for cognitive radio *ad-hoc* network is discussed in [95]. The scheme utilizes single transceiver with a common control channel. All nodes in the network switch to control channel for the purpose of contention and the winning pair switches to the data channel for the purpose of communication. Before performing communication, the data channel is sensed for primary incumbent activity by the communicating pair during quiet-period. If the channel is sensed busy, the communication is dropped for the current superframe. A superframe structure of OSA-MAC is shown in Fig. 1.9. The protocol is able to solve the problem of multi-channel hidden terminal through temporal synchronization during contention phase.

In [96], the concept of busy-tone is introduced, where a small band of spectrum is mapped to a data channel. This band of spectrum is termed as busy-tone-band, and is responsible for identifying whether a data channel corresponding to the busy-tone-band is occupied or not. Whenever a user occupies a data channel it sends a busy tone to its corresponding busy-tone-band. A common control channel is utilized for control signals handshake. Three transceivers are utilized, each assigned on data channel,

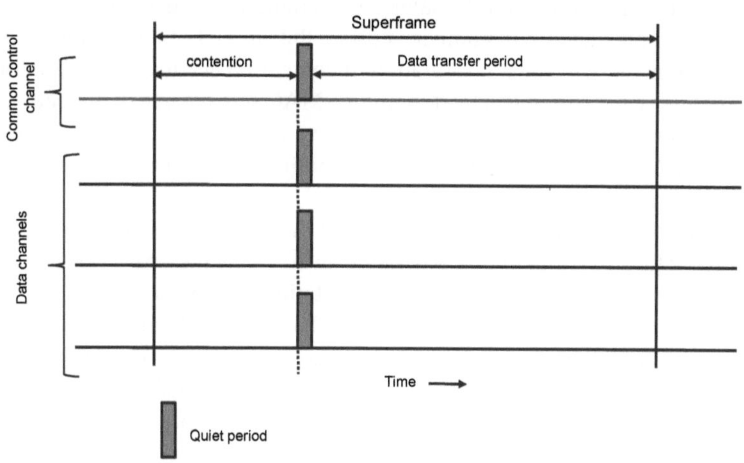

Fig. 1.9 Superframe structure of OSA-MAC. All nodes switch to the common control channel during the contention period. The winning node pair move to the agreed data channel and perform sensing of the channel during quiet-period. Based on the sensing result, the node pair either defer their communication for the current superframe or go ahead with the communication

control channel and busy-tone-band. This protocol also solves the problem of hidden terminal. A multi-transceiver based MAC is also discussed in [97], which utilizes a dedicated transceiver for a dedicated common control channel. The scheme follows distributed channel assignment by following a list of currently used bands and free bands. The sender and receiver follow data exchange similar to optional RTS-CTS IEEE 802.11 standard on a common agreed available channel. In [98], a distributed cognitive radio MAC protocol has been discussed which provides detection and protection of incumbent systems around the cognitive radio user pair. The literature also describes a method for sensing information exchange between neighbor nodes. In [99], an adaptive MAC protocol for throughput enhancement in cognitive radio has been discussed which changes its transmission mode between frame recovery mode and dual transmit mode according to the channel status. It utilizes two half-duplex transceivers to achieve this goal. The CR MAC protocol proposed in [100] utilizes multiple channel sensor and CR-enabled transceiver for efficient utilization of the unused spectrum while avoiding collision with PUs. It closely integrates the cooperative sequential spectrum sensing along with packet scheduling at MAC layer. A Markov chain model is discussed to study the network in saturated as well as in non-saturated condition.

1.7.2 Time-Slotted Based MAC Scheme

The utilization of random access technique simplifies the design of access scheme but decreases the efficiency of overall throughput of the system. Also the requirement of contention makes the control channel vulnerable to control channel saturation problem. With the increase in number of users, the amount of contention on control channel will increase leading to saturation of contention and hence the data transfer. Such problems can be removed by designing a MAC scheme which does not involve contention on a common control channel or by removing contention completely. Time-slotted MAC protocols remove the contention but require network wide tight synchronization of time.

IEEE WRAN 802.22 was briefly discussed in Sect. 1.4.1, which is the standard for infrastructure based cognitive operation over TV white space so as to enable broadband access in rural areas [59]. A base station coordinates all its users which are termed as consumer premise equipment (CPE). The MAC scheme is time-slotted, comprising of superframes, which in turn comprises of multiple MAC frames. Superframe control header (SCH) precedes all superframe and consists of network information which is distributed to CPEs. Each MAC frame is composed of two subframes: *Downstream* (DS) and *Upstream* (US). A DS subframe consists of a single burst of packet from a given CPE along with the scheduling information of user bursts. A US subframe consists of multiple packet bursts, each transmitted from different CPE. Among others, a US subframe is also used for self-coexistence through the help of coexistence beacon protocol (CBP). CBP is discussed in Sect. 1.8. The standard supports two type of spectrum sensing: *fast* and *fine*, allowing control over latency and quality of spectrum sensing results respectively. In [101], an improvement to IEEE 802.22 standard has been proposed. The discussed centralized MAC protocol for cognitive user coexists with primary user by using in-band/out-band signalling mechanism and by utilizing extra channels to broadcast current channel list.

In [102], a cognitive MAC protocol for multi-channel wireless (C-MAC) is proposed which utilizes distributed quiet period for primary user (PU) signal detection. In this protocol each channel is logically divided into recurring superframes comprising slotted beaconing period in which users exchange information and performs channel contention. The concept of rendezvous channel (RC) is introduced in which users synchronize, coordinate, perform reservation of channels. Users join a network through the help of beacon period of the superframes of a channel. The concept of distributed beaconing is utilized for solving the hidden terminal problem, by extending the beacon through re-broadcast. The utilization of multiple transceiver allows the MAC scheme to be robust against spectrum changes, as out-of-band sensing is performed on distributed quiet periods. The MAC scheme is targeted towards providing higher aggregate link throughput.

A combination of time-slotted and random access schemes have emerged as a new design trend for MAC schemes. Some literatures refer to them as hybrid MAC protocols [11]. In [103], a single transceiver based MAC scheme is proposed which utilizes a dedicated common control channel for signaling purposes. The protocol utilizes pre-defined window period for user coordination over the choice of frequency spectrum. Users form different cluster, with each cluster member having the intention to communicate within the cluster. The communication is done by utilizing conventional contention based data transfer. In [104], a control channel-less MAC protocol has been proposed and referred as Synchronized MAC (SYN-MAC) protocol for multi-hop cognitive radio networks. In this approach all nodes are assumed to be equipped with two radios. One radio is used for listening (listening radio) control signals and the other for both receiving and transmitting data (data radio). Here time is divided into time-slots, where each slot represents a data channel. These time-slots are used for control signal exchange, while data exchange can be done on any agreed channel. Users switch to the channels during the start of the time-slot and if interested perform control signal exchange. An opportunistic MAC scheme with the requirement of two transceivers and a dedicated common control channel is proposed in [105]. One radio is tuned to CCC while the other is tuned to data channels. The data channels are time-slotted while CCC is partly slotted with random access duration. During the partly slotted duration of CCC, users communicate about the availability of channels through sensing. During the random access period, users perform negotiation of channel access through contention. In [106], a distributed sensing mechanism has been proposed utilizing two radios; one radio is used for communication purpose while the other one is used for dedicated sensing purpose. Therefore it allows simultaneous communication and sensing. A dedicated CCC is utilized for signaling purposes. The time is divided into fixed length beacon intervals with two phases: *Announcement Traffic Indication Message* (ATIM) window and *DATA* window. Minislot protocol is utilized during ATIM window to learn and dissipate network-wide spectral opportunities. The actual communication is performed during DATA window. [107, 108] apply the multi-channel and Time Division Multiple Access (TDMA) technique to increase the throughput of the cognitive radio network. The system is divided into fixed length time frames comprising of ATIM window and communication window similar to that discussed in [56]. The communication window is further divided into fixed size time-slots for the provision of Time Division Multiple Access. The contention is done during the ATIM window and it reserves the channel-time slots for actual transmission during the communication window.

1.8 Self-coexistence in Cognitive Radio Networks

With multiple cognitive radio networks operating in a given region, a node in a network has to mitigate the harmful effects of interference from licensed users and nodes of other cognitive radio networks. Interference from other cognitive radio network involves: different cognitive radio network operating with different set of protocols (heterogeneous) or different cognitive radio networks operating under the same set of protocol (homogeneous). Mitigating interference from cognitive radio network operating under same set of protocol requires self-coexistence techniques. Self-coexistence techniques are designed so as two or more homogeneous cognitive radio network can cooperate and coexist with minimal interference.

A three step self-coexistence technique is proposed in [109]. Each of these steps provide self-coexistence from different perspective. A sequential flow is presented for making a decision on the selection of technique, which includes parameters like coexistence complexity, reconfigurability and performance of the selected scheme. The techniques in the order of selection process are: *resource relocation*, *transmission power control*, and *resource sharing*.

1. **Resource Relocation**: Resource in the context of cognitive radio network refers to a channel or a combination of time-duration with a channel [109]. The first step in resource relocation is to identify the resources over which the users are interfering. Once these resources are identified, they are redistributed among the users such that they do not interfere while also maintaining a certain level of QoS for their respective network. Resource relocation achieves high performance gain in comparison to other methods of self-coexistence. Also the complexity of this method is lower than other methods of self-coexistence.
2. **Transmission Power Control** (TPC): This technique is performed when resource relocation based self-coexistence is not feasible. This technique requires adapting the transmission power of interfered users so as to separate the interference range of users. TPC technique allows self-coexistence but at the cost of decrease in communication range and throughput. Therefore an optimization is required to negotiate between the level of interference and QoS requirement.

3. **Resource Sharing**: When resource relocation and TPC based self-coexistence are not successful, then the resource is shared in frequency or time domain between the interfered users. Sharing a spectrum in frequency domain implies dividing a large block of spectrum into small chunks (fragmentation) and allocating them to the interfered users [110]. Similarly sharing a spectrum in time-domain implies sharing it on time division multiplex basis, where a particular time duration is allocated to a given interfered user [111]. This technique is more complex and less efficient compared to other techniques of self-coexistence.

1.8.1 Resource Relocation Based Self-coexistence

Various works on resource relocation have been done in wireless networks from the perspective of self-coexistence [112]. These methods can be applied in cognitive radio networks with the additional design consideration of sporadically available channels. In IEEE 802.22 standard, coexistence beacon protocol is utilized for supporting intra-network coexistence [113]. Intra-network communication is done with the help of CBP [114] where intra-network packets are transmitted during a specifically defined duration: coexistence window. During this duration, contention based access is done to access the spectrum with higher priority given to the inter-base station messages.

Since resource relocation implies resource allocation, graph coloring based channel allocation schemes can be utilized for the same [115]. The users in the network are mapped as vertices while interference between them is mapped as edges between vertices. Each user has a list of available colors (channels). The aim of a graph coloring scheme is to allocate colors to vertices such that no two neighbor vertices are allocated with the same color. In infrastructure based network, base stations are mapped as vertices. In [116], a graph coloring based channel allocation scheme is proposed which allocates channel to different users in a cognitive radio network with proportional fairness. A collaborative labeling rule is proposed which guarantees users' fairness of service capacity. In [117], a utility graph coloring scheme is proposed which consider allocation of colors based on the weights assigned to them. The overall aim of the technique is to maximize the overall spectrum utilization. An enhanced MAC scheme for IEEE 802.22 standard is also proposed to maximize the spectral efficiency with the help of channel aggregation and fragmentation. In [118], the channel allocation problem is converted into multi-stage modified minority game, where the aim is to select a channel free from interference. The number of competitors is known with the help of foreign beacon period (broadcast beacons of foreign 802.22 networks) of IEEE 802.22 network. The player in the game makes a decision whether to stay on the current channel or to move to other channel with low interference. A channel switching cost function is defined for the same. A dynamic hopping based self-coexistence technique for IEEE 802.22 networks is proposed in [114]. A group of IEEE 802.22 networks form a community and hop periodically in a defined manner from one channel to another. The successive access of channel by different networks is separated by quiet-period, during which other networks can perform out-of-band sensing. This increases the performance and reliability of the system while maintaining coexistence among all IEEE 802.22 networks.

1.8.2 Resource Sharing Based Self-coexistence

Frequency division multiplexing (FDM) based coexistence has been widely studied over the years for allowing systems to coexist in orthogonal frequency domain [110]. It allows a wide band of spectrum to be sliced into small fragments of spectrum bands, which can be allocated to individual users. The division of wide frequency band into small bands results in fragmentation, which has been widely studied for dynamic spectrum access. In [119, 120], performance analysis of spectrum fragmentation is done. It is shown that excessive fragmentation increases the spectrum utilization but decreases the performance of the system. Experimental and analytical model have been proposed to validate the system. Since frequency division multiplexing is done at the Layer 1 (PHY) of the protocol stack, it is beyond the scope of this work. Readers are suggested to go through the literature work done in [121], which introduces time-frequency-code slicing technique for multi-rate support in networks.

Time division based multiplexing (TDM) has been widely used for resource distribution [111]. In TDM, the resource access is divided into small time durations, where all users are allocated with disjoint time-durations to access the given resource. The computing power distribution among multiple process in a computer system is an example of TDM system. Conventional *Global System for Mobile Communication* (GSM), is based on TDM, where multiple users are assigned the same band but are separated on the basis of TDM. GSM system utilizes FDM along with TDM to increase the number of users on the network. For self-coexistence purposes, [109] utilizes a TDM access (TDMA) based frame sharing approach. Here the superframe of each network for the same channel is divided in time-domain, such that each network has access to an exclusive non-overlapping superframe.

1.9 Discussion

The overcrowding of unlicensed bands has resulted in "tragedy of commons," that needs to be mitigated by allowing more unlicensed bands for unlicensed users. A paradox has been shown between the spectrum scarcity and its under-utilization, and therefore open spectrum access techniques have been envisioned by researchers to effectively utilize frequency spectrum. To realize the open spectrum sharing technique, underlay and overlay sharing techniques have been proposed. Underlay communication is useful for wide-band short range devices while overlay sharing is useful for opportunistic utilization of spectrum with region level transmission range. To realize the implementation of opportunistic utilization of spectrum, cognitive concept of software defined radio is introduced. A cognitive radio is a reconfigurable radio that can adapt its transmission characteristic according to the environment in which it operates. Several standards are being developed by various researchers across the world for the purpose of realization of open spectrum system.

The concept of quality of service from the perspective of throughput and latency is introduced. Latency guarantee in a network requires prioritization of a packet from the perspective of access time or queuing delay. While throughput guarantee in a network requires reservation of a particular resource (bandwidth), so that the system operates with acceptable grade of service. The performance of a network depends upon the channel on which it operates. Several research work on single-channel or multi-channel selection techniques have been introduced while making a spectrum decision on the basis of various spectrum characteristics. Channel selection in *infrastructure* and *ad-hoc* networks are also discussed, which differ in terms of distributed and collaborative behavior. To mitigate the access collision on a channel, MAC schemes have been introduced. These MAC schemes are classified into two types: random access based or time-slotted based. Random access schemes are generally utilized in *ad-hoc* network while time-slotted access schemes are generally utilized in *infrastructure* based network due to the requirement of tight-synchronization. To get the best of both classes of access protocols, a new class of MAC scheme has been introduced which is called as *hybrid* scheme. To avoid the mutual interference from other homogeneous CR networks in the vicinity, self-coexistence techniques have been discussed. Self-coexistence can be done from three perspective: resource relocation, transmission power control and resource sharing, in the order of increasing complexity and decreasing performance. The different technologies responsible for individual attribute of QoS can be integrated together to form a framework which allows QoS provisioning to the whole system.

References

1. Tragos, E., Zeadally, S., Fragkiadakis, A., & Siris, V. (2013). Pectrum assignment in cognitive radio networks: A comprehensive survey. *IEEE Communications Surveys Tutorials, 15*, 1108–1135.
2. Force, S. P. T. (2002). *Report of the spectrum rights and responsibilities working group.* Technical Report: Federal Communications Commission.
3. Akyildiz, I. F., Lee, W.-Y., Vuran, M. C., & Mohanty, S. (2006). Next generation/dynamic spectrum access/cognitive radio wireless networks: A survey. *Computer Networks, 50*, 2127–2159.
4. Lee, W.-Y., & Akyldiz, I. (2011). A spectrum decision framework for cognitive radio networks. *IEEE Transactions on Mobile Computing, 10*, 161–174.
5. Akyildiz, I. F., Lee, W.-Y., & Chowdhury, K. R. (2009). Crahns: Cognitive radio ad hoc networks. *ad hoc Networks, 7*, 810–836.
6. Ghasemi, A., & Sousa, E. (2005). Collaborative spectrum sensing for opportunistic access in fading environments. In *2005 First IEEE International Symposium on New Frontiers in Dynamic Spectrum Access Networks, 2005. DySPAN 2005* (pp. 131–136).
7. Hillenbrand, J., Weiss, T., & Jondral, F. (2005). Calculation of detection and false alarm probabilities in spectrum pooling systems. *IEEE Communications Letters, 9*, 349–351.

8. Ko, G., Franklin, A., You, S.-J., Pak, J.-S., Song, M.-S., & Kim, C.-J. (2010). Channel management in IEEE 802.22 wran systems. *IEEE Communications Magazine*, *48*, 88–94.
9. Na, D.-H., Nan, H., & Yoo, S.-J. (2007). Policy-based dynamic channel selection architecture for cognitive radio networks. In *Second International Conference on Communications and Networking in China, 2007. CHINACOM '07* (pp. 1190–1194).
10. Xu, D., Jung, E., & Liu, X. (2008). Optimal bandwidth selection in multi-channel cognitive radio networks: How much is too much?. In *3rd IEEE Symposium on New Frontiers in Dynamic Spectrum Access Networks, 2008. DySPAN 2008* (pp. 1–11).
11. Cormio, C., & Chowdhury, K. R. (2009). A survey on MAC protocols for cognitive radio networks. *ad hoc Networks*, *7*(7), 1315–1329.
12. Park, S. (2006). Diffserv quality of service support for multimedia applications in broadband access networks. In *International Conference on Hybrid Information Technology, 2006. ICHIT '06* (vol. 2, pp. 513–518).
13. Borosa, T., Marsic, B., & Pocuca, S. (2003). QoS support in ip multimedia subsystem using diffserv. In *Proceedings of the 7th International Conference on Telecommunications, 2003. ConTEL 2003* (vol. 2, pp. 669–672).
14. Wi-fi certified for wmm- support for multimedia applications with quality of service in wi-fi networks, (2004). *Technical Report, Wi-Fi Alliance.*
15. Dai, Q. (2011). A survey of quality of experience. *Proceedings of the 17th International Conference on Energy-aware Communications, EUNICE'11* (pp. 146–156). Berlin, Heidelberg: Springer.
16. Fall, K. R., & Stevens, W. R. (2011). *TCP/IP illustrated: The protocols* (Vol. 1). Boston: Addison-Wesley.
17. Telecom, I. (1996) Standardization sector of ITU. In *Video Coding for Low Bitrate Communication, Draft ITU-T Recommendation H* (vol. 263).
18. Commission, F. C., et al. (2003). Revision of parts 2 and 15 of the commission rules to permit unlicensed national information infrastructure (U-NII) devices in the 5 GHz band. *Memorandum Opinion and Order*, *21*.
19. Berger, R. J. (2003). Open spectrum: A path to ubiquitous connectivity. *ACM Queue*, *1*(3), 60–68.
20. Werbach, K. (2002) Open spectrum: The new wireless paradigm. *New America Foundation Spectrum Policy Program.*
21. Valletti, T. M. (2001). Spectrum trading. *Telecommunications Policy*, *25*(10), 655–670.
22. Li, H., Shu, T., He, F., & Song, J. B. (2013). Futures market for spectrum trade in wireless communications: Modeling, pricing and hedging. In *2013 IEEE Global Communications Conference (GLOBECOM)* (pp. 902–907).
23. Spectrum management handbook, (2014). Infocomm Development Authority of Singapore, Issue 1 Rev 2.4.
24. Hazlett, T. W. (2005). Spectrum tragedies. *Yale Journal on Regulation*, *22*, 242.
25. Kolodzy, P. J. (2006). Interference temperature: a metric for dynamic spectrum utilization. *International Journal of Network Management*, *16*(2), 103–113.
26. Aiello, G. R., & Rogerson, G. D. (2003). Ultra-wideband wireless systems. *IEEE Microwave Magazine*, *4*(2), 36–47.
27. Mitola, J., & Maguire, G. Q, Jr. (1999). Cognitive radio: Making software radios more personal. *IEEE Personal Communications*, *6*, 13–18.
28. Mitola, J. (1999) Cognitive radio for flexible mobile multimedia communications. In *1999 IEEE International Workshop on Mobile Multimedia Communications, 1999. (MoMuC '99)* (pp. 3–10).
29. Rondeau, T., & Bostian, C. (2009). *Artificial Intelligence in Wireless Communications*. Artech House: Artech House mobile communications series.
30. Islam, M., Koh, C., Oh, S. W., Qing, X., Lai, Y., Wang, C. et. al., (2008). Spectrum survey in singapore: Occupancy measurements and analyses. In *3rd International Conference on Cognitive Radio Oriented Wireless Networks and Communications, 2008. CrownCom 2008* (pp. 1–7).

31. Erpek, T., Lofquist, M., & Patton, K. (2007). Spectrum occupancy measurements loring commerce centre limestone, maine. *Shared Spectrum Company Report*. Retrieved September 18–20, 2007.
32. Erpek, T., Steadman, K., & Jones, D. (2007). Dublin ireland spectrum occupancy measurements collected. *Shared Spectrum Company Report*. Retrieved April 16–18, 2007.
33. McHenry, M. A., McCloskey, D., Roberson, D., & MacDonald, J. T. (2005, December). Spectrum occupancy measurements chicago, illinois. *Shared Spectrum Company Report*. Retrieved November 16–18, 2005.
34. McHenry, M. A., & Steadman, K. (2005, August). Spectrum occupancy measurements location 1 of 6: Riverbend park, great falls, virginia. *Shared Spectrum Company Report*. Retrieved April 7, 2004.
35. McHenry, M. A., & Steadman, K. (2005, August). Spectrum occupancy measurements location 2 of 6: Tyson's square center, vienna, virginia. *Shared Spectrum Company Report*. Retrieved April 9, 2004.
36. McHenry, M. A., & Chunduri, S. (2005, August). Spectrum occupancy measurements location 3 of 6: National science foundation building roof. *Shared Spectrum Company Report, Revision, 2*. Retrieved April 16, 2004.
37. McHenry, M. A., & McCloskey, D. (2005, August). Spectrum occupancy measurements location 4 of 6: Republican national convention, New York city. *Shared Spectrum Company Report, Revision, 2*. Retrieved New york August 30, 2004–September 3, 2004.
38. McHenry, M. A., & Steadman, K. (2005, August). Spectrum occupancy measurements location 5 of 6: National radio astronomy observatory (NRAO) green bank, west virginia. *Shared Spectrum Company Report, Revision, 3*. Retrieved October 10–11, 2004.
39. McHenry, M. A., & Steadman, K. (2005, August). Spectrum occupancy measurements: Shared spectrum building roof, vienna, virginia. *Shared Spectrum Company Report*. Retrieved December 15–16, 2004.
40. Valenta, V., Marslek, R., Baudoin, G., Villegas, M., Suarez, M., & Robert, F. (2010). Survey on spectrum utilization in Europe: Measurements, analyses and observations. In *2010 Proceedings of the Fifth International Conference on Cognitive Radio Oriented Wireless Networks Communications (CROWNCOM)* (pp. 1–5).
41. Patil, K., Skouby, K., Chandra, A., & Prasad, R. (2011). Spectrum occupancy statistics in the context of cognitive radio. In *2011 14th International Symposium on Wireless Personal Multimedia Communications (WPMC)* (pp. 1–5).
42. Sanders, F. (1998). Broadband spectrum surveys in denver, co, san diego, ca, and los angeles, ca: Methodology, analysis, and comparative results. In *1998 IEEE International Symposium on Electromagnetic Compatibility* (vol. 2, pp. 988–993).
43. Chiang, R., Rowe, G., & Sowerby, K. (2007). A quantitative analysis of spectral occupancy measurements for cognitive radio. In *IEEE 65th Vehicular Technology Conference, 2007. VTC2007-Spring* (pp. 3016–3020).
44. Wellens, M., Wu, J., & Mahonen, P. (2007). Evaluation of spectrum occupancy in indoor and outdoor scenario in the context of cognitive radio. In *2nd International Conference on Cognitive Radio Oriented Wireless Networks and Communications, 2007. CrownCom 2007* (pp. 420–427).
45. Yin, S., Chen, D., Zhang, Q., Liu, M., & Li, S. (2012). Mining spectrum usage data: A large-scale spectrum measurement study. *IEEE Transactions on Mobile Computing, 11*, 1033–1046.
46. Lopez-Benitez, M., Umbert, A., & Casadevall, F. (2009). Evaluation of spectrum occupancy in spain for cognitive radio applications. In *IEEE 69th Vehicular Technology Conference, 2009. VTC Spring 2009* (pp. 1–5).
47. Bao, V. N. Q., Cuong, L. Q., Phu, L. Q., Thuan, T. D., Quy, N. T., & Trung, L. M. (2011). Vietnam spectrum occupancy measurements and analysis for cognitive radio applications. In *2011 International Conference on Advanced Technologies for Communications (ATC)* (pp. 135–143).
48. Mitola, J. (2000). *Cognitive Radio—An Integrated Agent Architecture for Software Defined Radio*. DTech thesis, Royal Institute of Technology (KTH).

49. Mino Diaz, E., Gelpi, P., von Hafen, J., Hiebinger, L., Jamsa, T., Malmgren, G., et al., (2004). The winner project: Research for new radio interfaces for better mobile services (Special section wide band systems). In *IEICE Transactions on Fundamentals of Electronics, Communications and Computer Sciences* (vol. 87, pp. 2592–2598).
50. Mohr, W. (2005). The winner (wireless world initiative new radio) project-development of a radio interface for systems beyond 3G. In *IEEE 16th International Symposium on Personal, Indoor and Mobile Radio Communications, 2005. PIMRC 2005* (vol. 2, pp. 857–862).
51. Group, D. X. W. et al., (2003). The XG Vision, Request for Comments, Version 2.0. Prepared by BBN Technologies Cambridge, Massachusetts.
52. Group, D. X. W. et al., The XG architectural framework. request for comments. Prepared by BBN Technologies Cambridge, Massachusetts.
53. Harada, H., Alemseged, Y. D., & Holland, O. (2011). IEEE dynamic spectrum access networks (DYSPAN) standards committee. *2011 2nd International Conference on Wireless Communication, Vehicular Technology, Information Theory and Aerospace and Electronic Systems Technology (Wireless VITAE)* (pp. 1–5). New York: IEEE.
54. Tuttlebee, W. H. (2003). *Software Defined Radio: Enabling Technologies*. New Jersey: Wiley.
55. Pollin, S., Ergen, M., Timmers, M., Dejonghe, A., Van der Perre, L., Catthoor, F. et. al., (2006) Distributed cognitive coexistence of 802.15. 4 with 802.11. In *1st International Conference on Cognitive Radio Oriented Wireless Networks and Communications, 2006* (pp. 1–5). New York: IEEE.
56. IEEE 802.11 Working Group, (2012). Wireless Lan Medium Access control (MAC) and physical layer (PHY) Specifications. IEEE 802.11 Standard.
57. Mangold, S., Zhong, Z., Hiertz, G. R., & Walke, B. (2004). IEEE 802.11e/802.11k wireless lan: spectrum awareness for distributed resource sharing. *Wireless Communications and Mobile Computing*, *4*(8), 881–902.
58. Sherman, M., Mody, A., Martinez, R., Rodriguez, C., & Reddy, R. (2008). IEEE standards supporting cognitive radio and networks, dynamic spectrum access, and coexistence. *IEEE Communications Magazine*, *46*, 72–79.
59. Cordeiro, C., Challapali, K., Birru, D., & Sai Shankar, N. (2005). IEEE 802.22: the first worldwide wireless standard based on cognitive radios. In *2005 First IEEE International Symposium on New Frontiers in Dynamic Spectrum Access Networks, 2005. DySPAN 2005* (pp. 328–337).
60. Ansari, J., Zhang, X., Achtzehn, A., Petrova, M., & Mahonen, P. (2011). A flexible mac development framework for cognitive radio systems. In *2011 IEEE Wireless Communications and Networking Conference (WCNC)* (pp. 156–161).
61. Wysocki, T., & Jamalipour, A. (2009). Mac framework for intermittently connected cognitive radio networks. In *IEEE 20th International Symposium on Personal, Indoor and Mobile Radio Communications, 2009* (pp. 481–485).
62. Akin, S., & Gursoy, M. (2011). Performance analysis of cognitive radio systems under QoS constraints and channel uncertainty. *IEEE Transactions on Wireless Communications*, *10*, 2883–2895.
63. Ishibashi, B., Bouabdallah, N., & Boutaba, R. (2008). QoS performance analysis of cognitive radio-based virtual wireless networks. In *The 27th Conference on Computer Communications. INFOCOM 2008*. New York: IEEE
64. Liang, Z., Feng, S., Zhao, D., & Shen, X. (2011). Delay performance analysis for supporting real-time traffic in a cognitive radio sensor network. *IEEE Transactions on Wireless Communications*, *10*, 325–335.
65. Zhang, J., Chen, C., Xiang, Y., Zhou, W., & Vasilakos, A. (2013). An effective network traffic classification method with unknown flow detection. *IEEE Transactions on Network and Service Management*, *10*, 133–147.
66. Lee, H.-W., & Chong, S. (2007). Combined packet scheduling and call admission control with minimum throughput guarantee in wireless networks. *IEEE Transactions on Wireless Communications*, *6*, 3080–3089.

67. Pattara-Aukom, W., Banerjee, S., & Krishnamurthy, P. (2002) Starvation prevention and quality of service in wireless lans. In *The 5th International Symposium on Wireless Personal Multimedia Communications, 2002* (vol. 3, pp. 1078–1082).
68. Sakurai, T., & Vu, H. (2007). Mac access delay of IEEE 802.11 dcf. *IEEE Transactions on Wireless Communications, 6*, 1702–1710.
69. Bianchi, G. (1998). Ieee 802.11-saturation throughput analysis. *IEEE Communications Letters, 2*, 318–320.
70. Ji, L., Arvanitis, T., & Woolley, S. (2003). Fair weighted round robin scheduling scheme for diffserv networks. *Electronics Letters, 39*(2), 333–335.
71. Perez-Costa, X., & Camps-Mur, D. (2010). IEEE 802.11e QoS and power saving features overview and analysis of combined performance (accepted from open call). *IEEE Wireless Communications, 17*, 88–96.
72. Choi, S., del Prado, J., Sai Shankar, N., & Mangold, S. (2003). IEEE 802.11 e contention-based channel access (EDCF) performance evaluation. In *IEEE International Conference on Communications, 2003. ICC '03* (vol. 2, pp. 1151–1156).
73. Gamal, A. Mammen, J., Prabhakar, B., & Shah, D. (2004). Throughput-delay trade-off in wireless networks. In *Twenty-third Annual Joint Conference of the IEEE Computer and Communications Societies. INFOCOM 2004* (vol. 1, p. 475).
74. Kaneko, M., Popovski, P., & Hayashi, K. (2009). Throughput-guaranteed resource-allocation algorithms for relay-aided cellular ofdma system. *IEEE Transactions on Vehicular Technology, 58*, 1951–1964.
75. Eryilmaz, A., & Srikant, R. (2007). Fair resource allocation in wireless networks using queue-length-based scheduling and congestion control. *IEEE/ACM Transactions on Networking, 15*, 1333–1344.
76. Barry, M., Campbell, M., Veres, A. (2001) Distributed control algorithms for service differentiation in wireless packet networks. In *Proceedings of the Twentieth Annual Joint Conference of the IEEE Computer and Communications Societies. INFOCOM 2001* (vol. 1, pp. 582–590). New Jersey: IEEE.
77. Joe, I., & Son, S. (2008). Dynamic spectrum allocation mac protocol based on cognitive radio for QoS support. In *Japan-China Joint Workshop on Frontier of Computer Science and Technology, 2008. FCST '08* (pp. 24–29).
78. Masonta, M., Mzyece, M., & Ntlatlapa, N. (2013). Spectrum decision in cognitive radio networks: A survey. *IEEE Communications Surveys Tutorials, 15*, 1088–1107.
79. Chandra, R., Mahajan, R., Moscibroda, T., Raghavendra, R., & Bahl, P. (2008). A case for adapting channel width in wireless networks. *SIGCOMM Computer Communivation Review, 38*, 135–146.
80. Bahl, P., Chandra, R., Moscibroda, T., Murty, R., & Welsh, M. (2009). "White space networking with wi-fi like connectivity. *Proceedings of the ACM SIGCOMM 2009 Conference on Data Communication, SIGCOMM '09* (pp. 27–38). New York: ACM.
81. Malanchini, I., Cesana, M., & Gatti, N. (2009). On spectrum selection games in cognitive radio networks. In *Global Telecommunications Conference, (2009). GLOBECOM 2009* (pp. 1–7). New York: IEEE.
82. Azarfar, A., Frigon, J.-F., & Sanso, B. (2012). Improving the reliability of wireless networks using cognitive radios. *IEEE Communications Surveys Tutorials, 14*, 338–354.
83. Yuan, Y., Bahl, P., Chandra, R., Moscibroda, T., & Wu, Y. (2007). Allocating dynamic time-spectrum blocks in cognitive radio networks. *Proceedings of the 8th ACM International Symposium on Mobile ad hoc Networking and Computing, MobiHoc '07* (pp. 130–139). New York: ACM.
84. Nie, N., Comaniciu, C. (2005). Adaptive channel allocation spectrum etiquette for cognitive radio networks. In *2005 First IEEE International Symposium on New Frontiers in Dynamic Spectrum Access Networks, 2005. DySPAN 2005* (pp. 269–278).
85. Song, Y., Fang, Y., & Zhang, Y. (2007). Stochastic channel selection in cognitive radio networks. In *Global Telecommunications Conference, 2007. GLOBECOM '07* (pp. 4878–4882). New York: IEEE.

86. Yang, G.-H., Zheng, H., Zhao, J., & Li, V.-K. (2006). Adaptive channel selection through collaborative sensing. In *IEEE International Conference on Communications, 2006. ICC '06* (vol. 8, pp. 3753–3758) .
87. Bennai, M., Sydor, J., & Rahman, M. (2010). Automatic channel selection for cognitive radio systems. In *2010 IEEE 21st International Symposium on Personal Indoor and Mobile Radio Communications (PIMRC)* (pp. 1831–1835).
88. Kung, L.-H. (2008). Channel selection for cognitive radio terminals, 2008. CS229 Project Report
89. Talat, S., & Wang, L.-C. (2009). QoS-guaranteed channel selection scheme for cognitive radio networks with variable channel bandwidths. In *International Conference on Communications, Circuits and Systems, 2009. ICCCAS 2009* (pp. 241–245).
90. Ververidis, C. N., Nasreddine, J., & Mähönen, P. (2009). Distributed channel selection protocol for single-hop transmissions in cognitive radio networks. *Proceedings of the 6th ACM Symposium on Performance Evaluation of Wireless ad hoc, Sensor, and Ubiquitous Networks, PE-WASUN '09* (pp. 155–156). New York: ACM.
91. Lien, S.-Y., Tseng, C.-C., & Chen, K.-C. (2008). Carrier sensing based multiple access protocols for cognitive radio networks. In *IEEE International Conference on Communications, 2008. ICC '08* (pp. 3208–3214).
92. Lo, B. F. (2011). A survey of common control channel design in cognitive radio networks. *Physical Communication*, *4*(1), 26–39.
93. Jia, J., Zhang, Q., & Shen, X. (2008). Hc-mac: A hardware-constrained cognitive MAC for efficient spectrum management. *IEEE Journal on Selected Areas in Communications*, *26*, 106–117.
94. Ma, L., Shen, C.-C., & Ryu, B. (2007). Single-radio adaptive channel algorithm for spectrum agile wireless ad hoc networks. In *2nd IEEE International Symposium on New Frontiers in Dynamic Spectrum Access Networks, 2007. DySPAN 2007* (pp. 547–558).
95. Le, L., & Hossain, E., (2008). A mac protocol for opportunistic spectrum access in cognitive radio networks. In *Wireless Communications and Networking Conference, 2008. WCNC 2008* (pp. 1426–1430). New York: IEEE.
96. Ma, L., Han, X., & Shen, C.-C. (2005). Dynamic open spectrum sharing mac protocol for wireless ad hoc networks. In *2005 First IEEE International Symposium on New Frontiers in Dynamic Spectrum Access Networks. DySPAN 2005* (pp. 203–213).
97. Pawelczak, P., Venkatesha Prasad, R., Xia, L., Niemegeers, I. G. M. M. (2005). Cognitive radio emergency networks - requirements and design. In *2005 First IEEE International Symposium on New Frontiers in Dynamic Spectrum Access Networks. DySPAN 2005* (pp. 601–606).
98. Yoo, S.-J., Nan, H., & Hyon, T.-I. (2009). Dcr-mac: distributed cognitive radio mac protocol for wireless ad hoc networks. *Wireless Communications and Mobile Computing*, *9*(5), 631–653.
99. Lee, B., & Rhee, S. H. (2008). Adaptive mac protocol for throughput enhancement in cognitive radio networks. In *International Conference on Information Networking, 2008. ICOIN 2008* (pp. 1–5).
100. Zhang, X., & Su, H. (2011). Cream-mac: Cognitive radio-enabled multi-channel mac protocol over dynamic spectrum access networks. *IEEE Journal of Selected Topics in Signal Processing*, *5*, 110–123.
101. Kim, H.-J., Jo, K.-J., Hyon, T.-I., Kim, J.-M., & Yoo, S.-J. (2008). Cognitive Radio Mac Protocol for Hidden Incumbent System Detection (vol. 5200, pp. 80–89).
102. Cordeiro, C., & Challapali, K. (2007). C-mac: A cognitive mac protocol for multi-channel wireless networks. In *2nd IEEE International Symposium on New Frontiers in Dynamic Spectrum Access Networks, 2007. DySPAN 2007* (pp. 147–157).
103. Hamdaoui, B., & Shin, K. (2008). Os-mac: An efficient mac protocol for spectrum-agile wireless networks. *IEEE Transactions on Mobile Computing*, *7*, 915–930.
104. Kondareddy, Y., & Agrawal, P. (2008). Synchronized mac protocol for multi-hop cognitive radio networks. In *IEEE International Conference on Communications, 2008. ICC '08* (pp. 3198–3202).

105. Su, H., & Zhang, X. (2007). Opportunistic mac protocols for cognitive radio based wireless networks. In *41st Annual Conference on Information Sciences and Systems, 2007. CISS '07* (pp. 363–368).
106. Timmers, M., Pollin, S., Dejonghe, A., Van der Perre, L., & Catthoor, F. (2010). A distributed multichannel mac protocol for multihop cognitive radio networks. *IEEE Transactions on Vehicular Technology, 59*, 446–459.
107. Ke, Y., & Yoo, S.-J. (2009). Mcr-mac: Multi-channel cognitive radio mac protocol for cooperative incumbent system protection in wireless ad-hoc network. In *First International Conference on Ubiquitous and Future Networks, 2009. ICUFN 2009* (pp. 6–11).
108. Kamruzzaman, S. M., Hamid, M. A., & Abdullah-Al-WADUD, M. (2010). An energy-efficient mac protocol for QoS provisioning in cognitive radio ad hoc network. *Radioengineering, 19*, 567–578.
109. Lee, D. H., Jeon, W. S., Jeong, D. G., & Ko, G. (2010). Self-coexistence techniques for cognitive radio lans/pans. In *2010 IEEE 21st International Symposium on Personal Indoor and Mobile Radio Communications (PIMRC)* (pp. 1516–1520).
110. Weinstein, S., & Ebert, P. (1971). Data transmission by frequency-division multiplexing using the discrete fourier transform. *IEEE Transactions on Communication Technology, 19*, 628–634.
111. Falconer, D., Adachi, F., & Gudmundson, B. (1995). Time division multiple access methods for wireless personal communications. *IEEE Communications Magazine, 33*, 50–57.
112. Li, W., Li, W., & Pan, Y. (2006). *Resource Allocation in Next Generation Wireless Networks. Wireless Networks and Mobile Computing*. New York: Nova Science Publishers.
113. Working Group on Wireless Regional Area Networks, IEEE std. 802.22 (2011). http://www.ieee802.org/22/.
114. Hu, W., Willkomm, D., Abusubaih, M., Gross, J., Vlantis, G., Gerla, M., et al. (2007). Cognitive radios for dynamic spectrum access - dynamic frequency hopping communities for efficient IEEE 802.22 operation. *IEEE Communications Magazine, 45*, 80–87.
115. Han, Z., & Liu, K. (2008). *Resource Allocation for Wireless Networks: Basics, Techniques, and Applications*. Cambridge: Cambridge University Press.
116. Ge, Y., Sun, J., Shao, S., Yang, L., & Zhu, H. (2010). An improved spectrum allocation algorithm based on proportional fairness in cognitive radio networks. In *2010 12th IEEE International Conference onCommunication Technology (ICCT)* (pp. 742–745).
117. Sengupta, S., Brahma, S., Chatterjee, M., & Shankar, N. S. (2013). Self-coexistence among interference-aware IEEE 802.22 networks with enhanced air-interface. *Pervasive and Mobile Computing, 9*(4), 454–471.
118. Sengupta, S., Chandramouli, R. Brahma, S. & Chatterjee, M. (2008). A game theoretic framework for distributed self-coexistence among IEEE 802.22 networks. In *IEEE Global Telecommunications Conference, 2008. IEEE GLOBECOM 2008* (pp. 1–6).
119. Coffman, E., Robert, P., Simatos, F., Tarumi, S., & Zussman, G. (2010). Channel fragmentation in dynamic spectrum access systems: A theoretical study. *Proceedings of the ACM SIGMETRICS International Conference on Measurement and Modeling of Computer Systems, SIGMETRICS '10* (pp. 333–344). New York: ACM.
120. Coffman, E., Robert, P., Simatos, F., Tarumi, S., & Zussman, G. (2012). A performance analysis of channel fragmentation in dynamic spectrum access systems. *Queueing Systems, 71*(3), 293–320.
121. Karol, M., Haas, Z., Woodworth, C., & Gitlin, R. (1997). Time-frequency-code slicing: efficiently allocating the communications spectrum to multirate users. *IEEE Transactions on Vehicular Technology, 46*, 818–826.

Chapter 2
Cognitive Radio Network- A Review

As discussed in Chap. 1, the quality of service and energy efficiency of a cognitive radio system depends upon various parameters, including spectrum selection, media access scheme and spectrum sensing order. In this chapter, we provide a review on technologies which facilitates/improves the parameters responsible for QoS and energy management. At the end of this chapter, we will discuss about different cognitive radio platforms and their evolution.

2.1 Spectrum Management

Channel selection (spectrum management) forms an integral part of quality of service framework for cognitive radio network. Since different channels have different local policy restriction, their usage offer different grade of services depending upon the operating frequency [1]. Channel usage behavior of incumbents can be characterized as deterministic or stochastic, leading a secondary user to utilize different etiquettes for efficient spectrum utilization.

2.1.1 Ant Colony Optimization Based Spectrum Management

A channel selection technique is discussed in [2] which enables each secondary user (SU) to detect appropriate channels over which it can communicate without any coordination with the other SUs. The overall aim is to increase the spectrum utilization in the system. The authors discussed a biologically-inspired Spectrum Sharing (BIOSS) algorithm based on the adaptive task allocation model of insect colonies. Here the problem of channel allocation is reduced to ant colony optimization. Table 2.1 shows the mapping between CRN and insect colony.

An ant performs efficiently if the suited tasks are allocated to them individually and are allowed to simultaneously complete the task with other ants. Thus the col-

V. Mishra et al., *QoS and Energy Management in Cognitive Radio Network*, Signals and Communication Technology, DOI 10.1007/978-3-319-45860-1_2

Table 2.1 A mapping between ant colony and cognitive radio

Insect colony	Cognitive radio network
Insect	Cognitive radio
Task	Available channel
Task associated stimuli (s)	Permissible power to channel (P_j)
Response threshold (θ)	Required transmission power (p_{ij})

Table 2.2 A mapping between ant colony and cognitive radio

Condition	Events
$P_j \gg p_{ij}$	Probability of cognitive radio i to select channel j increases
$P_j \ll p_{ij}$	Probability of cognitive radio i to select channel j decreases
For constant p_{ij}, if P_j increases	Probability of cognitive radio i to select channel j increases
For constant P_j, if p_{ij} increases	Probability of cognitive radio i to select channel j decreases

laborative behavior of the ants optimizes the overall job of ant colony. This division if labour is based on the response threshold θ of each ant, which is the probability of reacting to a task associated stimuli(s); whereas task stimuli(s) is the intensity of activator associated with the task. The task is performed iff $s \gg \theta$.

The probability of performing a task is given by:

$$T_\theta(s) = \frac{s^n}{s^n + \theta^n} \tag{2.1}$$

Here, $n > 1$ determines steepness of the threshold.

Each cognitive radio i in the system calculates the channel selection probability for every available channel c as:

$$T_{ij}^{csp} = \frac{P_j^n}{P_j^n + \alpha p_{ij}^n + \beta L_{ij}^n} \tag{2.2}$$

where, L_{ij}: learning factor to enable CR i to learn or forget channel j and α, β: positive constants.

The channels are selected on the basis of conditions as shown in Table 2.2.

The spectrum selection is done by Algorithm 2.1 which executes over each cognitive radio node to provide spectrum selection based on QoS requirements.

Algorithm 2.1: Algorithm for load balancing in a given category of channels

```
Determine available spectrum bands;
Estimate P_j values for every channel j;
Set learning factor (L_ij) value to be same for each channel and learning coefficients (ξ0, ξ1)
  to appropriate values [2].
Computes T^csp for all available channels.
Selects channels with max T^csp.
if Selected channel meets QoS requirements then
  | L_ij = L_ij − ξ0 if selected channel meets QoS requirements.
else
  | L_ij = L_ij + ξ1 if selected channel does not meet QoS requirements.
end
```

Some of the observation worth to note is that, as the interference increase, P_j for a channel decrease. The channel with small P_j due to the interference can be preferred by the cognitive radio node which needs low transmission power. This will enable the cognitive radio node to control the interference in the channels. Also, the spectrum sharing becomes more efficient as the power demand of CR increases which inturn would prefers the channel with more transmission power.

2.1.2 Non-linear Optimization Based Spectrum Management

In [3], the authors have considered a spectrum assignment problem for cognitive radio sensor network via a Mixed Integer Non-linear Programming problem formulation. The spectrum assignment problem is considered under coverage, interference, minimum data rate and power budget constraints. The problem is transformed to Binary Linear Programming problem and further reduced to random, greedy and two-stage (decoupled) algorithms.

The network model comprises of a set of N cognitive radio-enabled wireless sensor nodes distributed within a set C of cells and coexisting with PUs operating in an up-link network on the same band, and transmitting on each channel with power P_{PU}. Both systems operate on a band comprising a set of K channels, each with bandwidth $B_k, \forall k \in K$. Only one node is $active$ to report the measured data while others are in $sleep$ mode.

$|h^k_{n,SU}|^2$ is the channel gain between node n and the cognitive radio base station receiver on channel k. $|h^k_{PU}|^2$ is the the channel gain between the primary user transmitter and the secondary user base station receiver on channel k. N_o is the power spectral density of a single sided additive white Gaussian noise. S is the set of active nodes.

Each of the *active sensors* from different cells is assigned channels in order to minimize transmit power with the following constraints:

(i) Interference at primary network receivers is less than given threshold. That is, $P_n^k|h_{n,PU}^k|^2 \le I_{th}^k, \forall k \in K, \forall n \in N$, where P_n^k is the transmission power of node n on channel k, $|h_{n,PU}^k|^2$ is the channel gain between node n and the primary user base station receiver on channel k, and I_{th}^k is the interference threshold at PU on channel k.

(ii) Mutual exclusion of channels among cognitive radio sensor nodes (A channel k is used by at most one node). That is, $\sum_{n=1}^{N} Y_n^k \le 1, \forall k \in K$, where Y_n^k is a binary variable identifying whether a node n is assigned channel k.

(iii) Transmission power is limited by a power budget. That is, $\sum_{k=1}^{K} Y_n^k P_n^k \le P_n^{max}, \forall n \in N$, where P_n^{max} is the maximum allowed power of node n.

(iv) Secondary network must achieve a minimum data rate (for QoS). That is,

$$\sum_{k=1}^{K} Y_n^k.B_k.log_2(1 + P_n^k * \gamma_n^k) \ge R_{th} \tag{2.3}$$

where γ_n^k is the channel to interference noise ratio of node n on channel k, and R_{th} is the minimum rate required required at the cognitive radio base station receiver.

(v) A coverage constraint where the geographical area is divided into multiple cells, each covers a certain event, and each cell must be covered by one active sensor node. That is, $\sum_{n=1}^{N} A_n^c Y_n = 1, \forall c \in C$, where A_n^c is a binary variable identifying whether a cognitive radio sensor node n is assigned a channel k, and $Y_n = \sum_{k=1}^{K} Y_n^k, \forall n \in N$, indicates whether a node is active or not.

The problem of choosing *active* sensors and assigning channels to these sensors can be formulated as:

$$\begin{aligned}
&\min_{Y_n^k, P_n^k} \sum_{n=1}^{N}\sum_{k=1}^{K} Y_n^k P_n^k \\
\textbf{subject to: } & P_n^k|h_{n,PU}^k|^2 \le I_{th}^k, \forall k \in K, \forall n \in N, \\
&\sum_{n=1}^{N} Y_n^k \le 1, \forall k \in K, \\
&\sum_{k=1}^{K} Y_n^k P_n^k \le P_n^{max}, \forall n \in N, \\
&\sum_{k=1}^{K} Y_n^k.B_k.log_2(1 + P_n^k * \gamma_n^k) \ge R_{th}, \forall n \in N, \\
&\sum_{n=1}^{N} A_n^c Y_n = 1, \forall c \in C
\end{aligned} \tag{2.4}$$

where,

$$A_n^c = \begin{cases} 1, & \text{if sensor } n \text{ cover area } c, \\ 0, & \text{otherwise} \end{cases}$$

$$Y_n^k = \begin{cases} 1, & \text{if sensor } n \text{ is assigned channel } k, \\ 0, & \text{otherwise} \end{cases}$$

$$\gamma_n^k = \frac{|h_{n,SU}^k|^2}{N_o * B + P_{PU} * |h_{PU}^k|^2}$$

The problem in Eq. 2.4 being NP-hard cannot be solved in polynomial time. Since cognitive radio sensor nodes are resource constrained, a realistic assumption regarding single transceiver on each node can be made. This would reduce the above multi-integer linear program to binary linear program as shown in Eq. 2.5.

$$\begin{aligned} & \min_{Y_n^k, P_n^k} \sum_{n=1}^{N} \sum_{k=1}^{K} Y_n^k P_n^k \\ \textbf{subject to: } & P_n^k |h_{n,PU}^k|^2 \leq I_{th}^k, \forall k \in K, \forall n \in N, \\ & \sum_{n=1}^{N} Y_n^k \leq 1, \forall k \in K, \\ & P_n^k \leq P_n^{max}, \forall k \in K, \forall n \in N, \\ & B_k.log_2(1 + P_n^k * \gamma_n^k) \geq R_{th}, \forall k \in K, \forall n \in N, \\ & \sum_{n=1}^{N} A_n^c Y_n = 1, \forall c \in C \end{aligned} \tag{2.5}$$

Algorithm 2.2: Algorithm for calculation of P_n^k

Calculate γ_n^k, $|h_{n,PU}^k|^2, \forall k \in K, \forall n \in N$;
Calculate P_n^k:
for $n = 1$ **to** N **do**
    for $k = 1$ **to** K **do**
        $P_n^k = P_n^k|_{min} = \frac{2^{R_{th}/B_k - 1}}{\gamma_n^k}$;
        if $P_n^k|_{min} > min\left(\frac{I_{th}^k}{|h_{n,PU}^k|^2}\right)$ **then**
            $P_n^k \Leftarrow \infty$;
        end
    end
end

Using Algorithm 2.2, the transmission power of each node n over all channels available in the network can be calculated. This can be utilized to generate a matrix Q, which provides transmission power of cognitive radio nodes in each cell. This can be given as $Q = P^k_{n_c}, \forall n \in N, \forall c \in C, \forall k \in K$ and can be calculated as given in [3]. The minimization problem given in Eq. 2.5 is reduced to a simpler minimization problem (Eq. 2.6) by utilizing the matrix Q.

For given $P^k_n, \forall n, \forall k$

$$\min_{Y^k_n} \sum_{n=1}^{N} \sum_{k=1}^{K} Y^k_n P^k_n$$

$$\textbf{subject to:} \sum_{n=1}^{N} Y^k_n \leq 1, \forall k \in K \tag{2.6}$$

$$\sum_{n=1}^{N} A^c_n * Y_n = 1, \forall c \in C$$

The discussed spectrum selection scheme utilizes following algorithms to solve the optimization problem in Eq. 2.6.

2.1.2.1 Minimum Power Algorithm

The aim of the Minimum power algorithm is to jointly optimize the selection of nodes and channels in such a way that the total transmission power is minimized. The transmission powers P^k_n from Algorithm 2.2 is utilized in Eq. 2.6 as shown in Algorithm 2.3.

Algorithm 2.3: Minimum Power Algorithm

```
1 Calculate γ_n^k, |h_{n,PU}^k|^2, ∀k ∈ K, ∀n ∈ N
2 Calculate P_n^k from Algorithm 2.2;
3 Calculate matrix Q from P_n^k;
4 Node/Channel Selection:
5 Solve Eq. 2.6 with the values of Q;
```

Assuming total number of bits in a binary representation for each of the optimal variables to be 1, the overall complexity of this algorithm is $O((NK)^{3.5})$.

2.1.2.2 Random Algorithm

The random algorithm choses a random node out of each cell and then assigns channels to it randomly. This can be achieved by the method discussed in Algorithm 2.4.

Algorithm 2.4: Random Algorithm

Calculate γ_n^k, $|h_{n,PU}^k|^2, \forall k \in K, \forall n \in N$;
Calculate P_n^k from Algorithm 2.2;
Calculate matrix Q from P_n^k;
Node Selection:
for $c = 1$ **to** C **do**
  $S(c) \Leftarrow$ select a random node;
end
Channel Selection:
for $c = 1$ **to** C **do**
  $S(s)_k \Leftarrow$ select a random channel $k \in K$;
  $K \Leftarrow K - k$;
end

The complexity of this algorithm is given as $O(1)$ for random selection of node/channel.

2.1.2.3 Greedy Algorithm

The greedy algorithm performs search over the matrix of P_n^k obtained from Algorithm 2.2 to find the node/channel pairs that has minimum transmission power. This can be achieved via the following Algorithm 2.5.

Algorithm 2.5: Greedy Algorithm

Calculate γ_n^k, $|h_{n,PU}^k|^2, \forall k \in K, \forall n \in N$;
Calculate P_n^k from Algorithm 2.2;
Calculate matrix Q from P_n^k;
Node/Channel Selection:
for $c = 1$ **to** C **do**
  $S(c)_k = min(P_n^k), \forall n \in N \forall k \in K$;
  $Q \Leftarrow Q - Q(:, k)$;
  $Q \Leftarrow Q - Q(c, :)$;
  $K \Leftarrow K - k$;
end

The total complexity of this greedy algorithm is of $O(C(NK)^2)$.

2.1.2.4 Two-Stage (Decoupled) Algorithm

In this method, the problem of spectrum allocation is broken into two sub-problems: *node-selection* and *channel-assignment*. In node-selection problem, life of node is extended by selecting nodes with higher residual energy as cells, while channel-assignment problem is solved as a standard linear program.

Algorithm 2.6: Two-Stage Algorithm

Calculate γ_n^k, $|h_{n,PU}^k|^2, \forall k \in K, \forall n \in N$;
Calculate P_n^k from Algorithm 2.2;
Calculate matrix Q from P_n^k;
Node Selection:
for $c = 1$ **to** C **do**
  $S(c) = max(n_{E_{res}}), \forall n \in N_c$;
end
Channel Selection:
Solve Eq. 2.6 with Q_s as input.

The overall complexity of this two-stage scheme is $O((CN^2) + (CK)^{3.5})$.

2.1.3 Game Theory Based Spectrum Management

A solution to address the problem of efficient spectrum pricing in cognitive radio network is given in [4]. The inefficiency of the Nash equilibrium (equilibrium pricing adopted by primary service providers) is discussed, in the sense that the profit of the primary service providers is not maximized. Here an alternate solution to this problem is discussed.

The problem is modelled as oligopoly market, in which a small number of firms compete with each other non-cooperatively and independently to achieve their objectives while dominating the overall market. The objective is determined in terms of controlling the price of products/quantity, and such actions are influenced by the decisions of other firms too. To achieve an equilibrium pricing scheme Bertrand game model may be utilized. Similarly, pricing for primary users' spectrum can also be calculated in a similar way, which are willing to share with secondary users.

A wireless system with total number of primary services N is assumed in which all primary services operate over different frequency spectrum $F_i, \forall i \in N$. Each of the primary service here wants to sell a portion of the available spectrum F_i to secondary users for price p_i as shown in Fig. 2.1.

An adaptive modulation is utilized to dynamically adjust the transmission rate. The spectral efficiency k of transmission by a secondary user is given by Eq. 2.7.

$$k = log_2(1 + K\gamma), \text{ where } K = \frac{1.5}{ln\left(\frac{0.2}{BER^{tar}}\right)} \tag{2.7}$$

where γ is the SNR at the receiver and BER^{tar} is the target bit-error-rate.

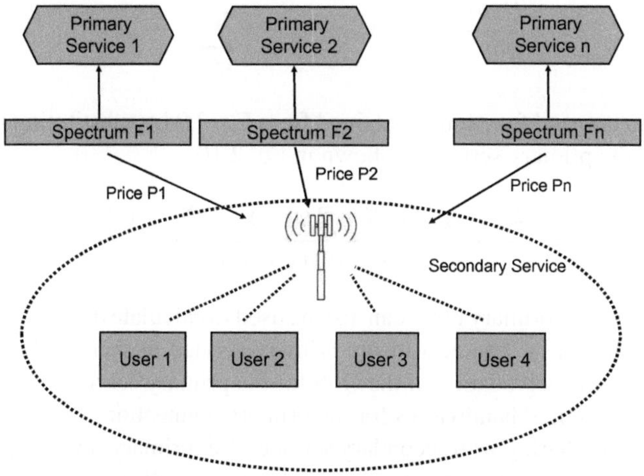

Fig. 2.1 Spectrum sharing model

Utility of the shared frequency spectrum is used to calculate the spectrum demand of the secondary users. A quadratic utility function is used to quantify the demand or utility of the spectrum.

$$U(b) = \sum_{i=1}^{N} b_i k_i^{(s)} - \frac{1}{2}\left(\sum_{i=1}^{N} b_i^2 + 2\nu \sum_{i \neq j} b_i b_j\right) - \sum_{i=1}^{N} p_i b_i \tag{2.8}$$

where b is the set consisting of the size of shared spectrum from all the primary services, i.e., $b = b_1, \ldots, b_N$; p_i is the price offered by primary service i, $k_i^{(s)}$ denotes the spectral efficiency of wireless transmission by a secondary user using primary users' frequency spectrum F_i. ν is the spectrum substitutability factor $\nu \in [-1.0, 1.0]$.

The quadratic utility function is used because of the following reasons:

(i) Quadratic function being concave, is able to represent the saturation of user satisfaction as the transmission rate increases.
(ii) A linear bandwidth demand function can be derived via differentiating the quadratic utility function. This eases the subsequent analysis for stability condition.
(iii) The spectrum substitutability factor ν, control the switching of secondary users onto its channels. If $\nu = 1$, a secondary user can switch freely among operating frequency spectra; while if $\nu = 0$, a secondary user can not switch freely.

The demand function for spectrum F_i is given by differentiating $U(b)$ with respect to b_i as shown in Eq. 2.9.

$$\frac{\partial U(b)}{\partial b_i} = 0 = k_i^{(s)} - b_i - v\sum_{i\neq j} b_j - p_i \tag{2.9}$$

The spectrum demand function is obtained from Eq. 2.9 by substituting the value of the prices of the primary service as shown in Eq. 2.10.

$$D_i(\mathbf{p}) = \frac{(k_i^{(s)} - p_i)(v(N-2)+1) - v\sum_{i\neq j}(k_j^{(s)} - p_j)}{(1-v)(v(N-1)+1)} \tag{2.10}$$

Price charged by the primary users can also be used to calculate the spectrum demand of the secondary users. The cost of the primary service in sharing the spectrum is computed from the degradation of the QoS for the primary service. The cost is only incurred if the required bandwidths for the primary connections cannot be provided due to bandwidth sharing with secondary services. For primary service i, the revenue function R_i and the cost function C_i, can be defined as follows:

$$\begin{aligned} R_i &= c_1 M_i \\ C_i(b_i) &= c_2 M_i \left(B_i^{req} - k_i^{(p)} \frac{W_i - b_i}{M_i} \right)^2 \end{aligned} \tag{2.11}$$

Here c_1 and c_2 denote the constant weight for revenue and cost functions. M_i is the number of primary connections, B_i^{req} is the bandwidth requirement and W_i is the size of spectrum of the the primary service i.

In the above given oligopoly market model, Bertrand game model can be utilized to formulate the game. Primary services act as the *players*, modifying the price per unit of the spectrum (p_i) act as the *strategy*, while the profit of sharing the spectrum with secondary services act as the *payoff*. Therefore the profit of each primary service can be calculated as shown in Eq. 2.12.

$$P_i(\mathbf{p}) = b_i p_i + R_i - C_i(b_i) \tag{2.12}$$

where $\mathbf{p}$ denotes the set of prices offered by all players in the game, that is $\mathbf{p} = \{p_1, p_2, p_3, \ldots, p_N\}$.

The best response function (best strategy of one player for given strategies of other players) of primary service i, given a set of prices offered by the other primary services $\mathbf{p}_{-i}$ is defined as follows:

$$B_i(\mathbf{p}_{-i}) = arg \max_{p_i} P_i(\mathbf{p}_{-i} \cup \{p_i\}) \tag{2.13}$$

An equilibrium is reached iff

$$p_i^* = B_i(\mathbf{p}_{-i}^*), \forall i \tag{2.14}$$

where $\mathbf{p}^* = \{p_1^*, p_2^*, \ldots, p_N^*\}$ is the set of prices when Nash equilibrium will be reached. $\mathbf{p}_{-i}$ is the set of best responses for player j such that $j \neq i$.

From Eqs. 2.12, 2.10 and 2.11, the profit function can be derived as shown in Eq. 2.15.

$$P_i(\mathbf{p}) = p_i D_i(\mathbf{p}) + c_1 M_i - c_2 M_i \left(B_i^{req} - k_i^{(p)} \frac{W_i - D_i(\mathbf{p})}{M_i} \right)^2 \tag{2.15}$$

The demand function in Eq. 2.10 can be expressed as: $D_i(\mathbf{p}) = D_1(\mathbf{p}_{-i}) - D_2 p_i$. Where $D_1(\mathbf{p}_{-i})$ and D_2 are given by Eq. 2.16.

$$\begin{aligned} D_1(\mathbf{p}_{-i}) =& \frac{k_i^{(s)}(v(N-2)+1) - v \sum_{i \neq j}(k_j^{(s)} - p_j)}{(1-v)(v(N-1)+1)} \\ D_2 =& \frac{(v(N-2)+1)}{(1-v)(v(N-1)+1)} \end{aligned} \tag{2.16}$$

For a Nash equilibrium to occur,

$$\frac{\partial P_i(p)}{\delta p_i} = 0, \tag{2.17}$$

From Eq. 2.15, 2.16 and 2.17, the Linear Eq. 2.18 can be obtained.

$$0 = 2c_2 k_i^{(p)} D_2 \left(B_i^{req} - k_i^{(p)} \frac{W_i - (D_1(\mathbf{p}_{-i}) - D_2 p_i)}{M_i} \right) + D_1(\mathbf{p}_{-i}) - 2D_2 p_i \tag{2.18}$$

For given parameters, the Eq. 2.18 provides the solution p_i^* which provides Nash equilibrium. From this the size of the shared spectrum can be calculated from Eq. 2.10 (i.e. $D_i(\mathbf{p}^*)$).

In reality, primary services may not be able to observe the profit gained by other primary services. In such scenario, for a Nash equilibrium to occur, a distributed form of price adjustment is required. Let the price offered by primary service i at iteration t be given as $p_i[t]$. If the strategies of the primary services in the previous iteration are observable by each other, then $p_i[t]$ is given as shown in Eq. 2.19.

$$p_i[t+1] = B_i(\mathbf{p}_{-i}[t]), \forall i \tag{2.19}$$

If the strategies used by other primary services are not known, then the relationship between the strategies in the current and the future iteration is to maximize the profit, which can be expressed as follows:

$$p_i[t+1] = p_i[t] + \alpha_i \left(\frac{\partial P_i(p)}{\partial p_i} \right) \tag{2.20}$$

where α_i is the learning rate.

The marginal profit of a primary service can be observed by the marginal spectrum demand for small variation in price ε as shown in Eqs. 2.21 and 2.22.

$$\frac{\partial P_i(\mathbf{p})}{\partial p_i} \approx \frac{P_i(\mathbf{p}_{-i}[t] \cup \{p_i[t] + \varepsilon\}) - P_i(\mathbf{p}_{-i}[t] \cup \{p_i[t] - \varepsilon\})}{2\varepsilon} \tag{2.21}$$

$$P_i(\mathbf{p}_{-i}[t] \cup \{p_i[t] \pm \varepsilon\}) = p_i D_i(\mathbf{p}_{-i}[t] \cup \{p_i[t] \pm \varepsilon\}) + c_1 M_i - c_2 M_i \left(B_i^{req} - k_i^{(p)} \frac{W_i - D_i(\mathbf{p}_{-i}[t] \cup \{p_i[t] \pm \varepsilon\})}{M_i} \right)^2 \tag{2.22}$$

To check if the above developed dynamic game will reach Nash equilibrium at the steady state, stability of the game is evaluated. Stability can be evaluated by utilizing the eigenvalues of the Jacobian matrix of a self-mapping function. The self mapping function is stable if the eigenvalues λ_i are inside the complex unit circle. Here the self mapping function utilized is from Eq. 2.20. Therefore the Jacobian matrix is defined as:

$$J = \begin{bmatrix} \frac{\partial p_1[t+1]}{\partial p_1[t]} & \frac{\partial p_1[t+1]}{\partial p_2[t]} & \cdots & \frac{\partial p_1[t+1]}{\partial p_N[t]} \\ \frac{\partial p_2[t+1]}{\partial p_1[t]} & \frac{\partial p_2[t+1]}{\partial p_2[t]} & \cdots & \frac{\partial p_2[t+1]}{\partial p_N[t]} \\ \vdots & \vdots & & \vdots \\ \frac{\partial p_N[t+1]}{\partial p_1[t]} & \frac{\partial p_N[t+1]}{\partial p_2[t]} & \cdots & \frac{\partial p_N[t+1]}{\partial p_N[t]} \end{bmatrix}$$

It can be observed that, if the strategies of other primary services are observable, then there is no control parameter to stabilize the price adaptation. If the strategies of other primary services are not observable, then the control parameter depends upon the learning rate.

For a repeated game, a trigger strategy allows a primary service to maintain collusion as long as other services agree to do so. In the case that the collusion is maintained forever, the long-term profit of primary service i can be expressed as follows:

$$P_i^0 + \delta_i P_i^0 + \delta_i^2 P_i^0 + \cdots = \frac{1}{1 - \delta_i} P_i^0$$

where P_i^0 is the profit due to the optimal price, and δ_i is the weight ($0 \leq \delta_i \leq 1$) which multiplies the profit in next stage.

For a game where one primary service deviates from the optimal price, it will experience the profit at the Nash equilibrium during the rest of the stages. The long term profit in this case can be expressed as:

$$P_i^d + \delta_i P_i^n + \delta_i^2 P_i^n + \cdots = P_i^d + \frac{\delta_i}{1 - \delta_i} P_i^n$$

where P_i^d and P_i^n are the profit due to price deviation and profit due to price at the Nash equilibrium respectively.

Collusion will be maintained if:

$$\frac{1}{1-\delta_i}P_i^0 \geq P_i^d + \frac{\delta_i}{1-\delta_i}P_i^n$$

2.1.4 Learning Automata Based Spectrum Selection

A spectrum selection scheme is discussed in [5], which minimizes the overall switching of the spectrum done by secondary users. This is done so as to mitigate the problem of aggravated delay and deteriorated packet-loss ratio which occurs due to constant channel switching of sporadically varying channel in cognitive radio environment. This is further complicated by the governing guarantee on maximum interference that can be caused to primary users by secondary users. Therefore a secondary user has to vacate a channel within a given time threshold whenever it detects the presence of primary users.

To minimize channel switching, the discussed stochastic channel selection algorithm utilizes learning automata which maximizes the number of successful transmissions. The algorithm asymptotically converges to an optimal channel which has higher availability in comparison to other channels.

In the given system, a secondary user can tolerate interference up to a threshold value; once the threshold value is reached, the secondary user has to leave the system. Whenever the channel is selected the secondary user transmits over consequent time slots, until the channel ceases to be free, Thus, a transmission is successful, if:

(i) The primary user is absent throughout the time slot.
(ii) Any of the neighbouring secondary user (the secondary users whose actions may potentially cause a transmission failure of the given secondary user) isn't operating at the same frequency.

The cognitive radio system is modelled as learning automata based control system, where the channel availability is modelled as random environment while the finite actions are modelled as the spectrum decisions. Each of these deterministic finite action will induce a random output from the random environment, which could be either favorable or unfavorable (i.e. whether selected channel has high availability or not). A learning automata algorithm can be utilized over this output to provide a control over the decision making of the finite actions. The learning automata algorithm essentially utilizes past results to make decisions about the next action.

During initialization of the secondary network, a secondary user selects any available channel at the beginning of each time slot for transmission and the output is denoted as β ($\beta = 0$ for successful and $\beta = 1$ for unsuccessful transmission). The system comprises of M primary users operating over C channels on which secondary users coexist in a hierarchical way. The system is viewed from the perspective of a secondary user denoted by k. The starting time of each slot is denoted by T_i. Each secondary user maintains a probability vector denoted by $\mathbf{P}_k = [p_1, \ldots, p_m]$ where p_i is probability of choosing channel i. Each secondary users also maintains an estimation vector denoted by $\mathbf{D}_k = [d_1, \ldots, d_m]$, where d_i is the estimated possibility of successful transmission by selecting channel i. $H(n)$ denotes the number of channels having higher values in the estimation vector D than the current selected channel. R and W are the resolution parameter and the initialization parameter respectively. δ represents the step size of adjusting the probability vector where, $\delta = 1/R$. $S_i(n)$ is the number of slots where the transmissions with selected channel i are successful, up to T_n. $C_i(n)$ is the number of slots where the channel i are selected to transmit packets, up to T_n.

The Algorithm 2.7 shows the learning automata based channel selection technique which executes over each node k with B as a predefined convergence threshold.

This channel selection technique is ε-optimal for any cognitive radio networks which is stationary. Therefore for any arbitrarily small $\varepsilon > 0$ and $\gamma > 0$, there exists a n' satisfying:

$$P_r\{|1 - p_z(n)| < \varepsilon\} > 1 - \gamma, \forall n > n'$$

where z is the index of the optimal channel in terms of probability of successful transmissions.

Algorithm 2.7: Algorithm for Stochastic Channel Selection

Initialize:
$P(0) = [p_1, \ldots, p_M]$ where $p_i = 1/M$ for all $1 \le i \le M$;
do
  Select available channel from $P(0)$;
  Evaluate and record β ;
while *Until each channel is selected W times*;
Update $S_i(0)$ and $C_i(0)$ according to value of βs for each channel i ;
Initializes $D(0) = [d_1, \ldots, d_M]$ where $d_i = S_i(0)/C_i(0)$;
do
  Select an available transmission channel i according to the probability vector $P(n)$;
  Update the probability vector $P(n)$ *as*:
  $p_j(n+1) = min\left(p_j(n) + \frac{\delta}{H(n)}, 1\right)\forall j$ if $d_j(n) > d_i(n)$
  $p_j(n+1) = max\left(p_j(n) - \frac{\delta}{M-H(n)}, 0\right)\forall j$ if $d_j(n) < d_i(n)$
  $p_i(n+1) = 1 - \sum_{j \ne i} p_j(n+1)$
  After T_n, the secondary user adjusts the value of D as:
  $S_i(n+1) = S_i(n) + (1 - \beta)$
  $C_i(n+1) = C_i(n) + 1$
  $d_i(n+1) = \frac{S_i(n+1)}{C_i(n+1)}$
while $(max(P(n)) > B)$;

2.1.5 Spectrum Selection in Varying Channel Bandwidth Environment

In [6], a scheme for cognitive radio spectrum hand-off is discussed where spectrum switching is done between channels of varying bandwidth. The concept of delay bandwidth product (DBP) is utilized to prioritize the channels with varying bandwidth. Here the delay bandwidth product is defined as the difference between the maximum tolerable delay threshold of the secondary user and the average occupation time of the primary user over a given channel. Based on this a secondary user can make spectrum selection to reduce power, increase range, increase throughput, load balance and guarantee its QoS requirement.

Theoretically delay bandwidth product refers to product of a data link's capacity and its round-trip delay time. Intuitively, this indicates the maximum amount of data that can exist in a network at any given time. Here the delay is considered as the difference between maximum tolerable delay of a secondary user and average occupation time of a primary user. As shown in Fig. 2.2, the total delay time (D_i) is the elapsed time until the secondary user gets the opportunity to transmit again. The total delay time is dependent on sensing time W_j, handoff time t_0 and the transmission time of secondary user T_k, where k is the current selected channel and j is candidate sensed channel. A secondary user can transmit again on the same selected channel or it can switch to another channel depending upon the DBP index.

The system comprises of N variable bandwidth channels each with bandwidth $B_i, \forall i \in (1, 2, \ldots, N)$. The current channel is defined as k while the candidate channel is defined as j. The total delay time D_i of secondary user i can be expressed as shown in Eq. 2.23.

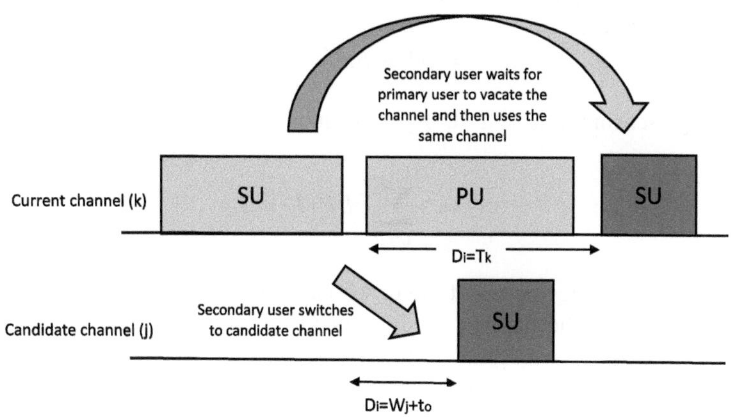

Fig. 2.2 Channel selection by secondary user

$$D_i = \begin{cases} t_0 + W_j, & \text{if } 1 \leq j \leq N, j \neq k; \\ T_k, & j = k \end{cases} \tag{2.23}$$

If the sensing time is less, the transmission time increases. Also for a higher bandwidth channel, there is an increase in throughput. Therefore an optimization is required between the bandwidth of the channel and the effective delay required by the channel itself. The values of T_k and W_j are dependent on what kind of traffic model is utilized in the system.

Two different primary user traffic models are considered for channel availability: *Pareto distribution model* and *Two-state Markov chain model.*

1. Pareto distribution model:
The probability density function (f(x)) and cumulative distribution function (F(x)) are described as shown in Eq. 2.24.

$$\begin{aligned} f(x) &= \frac{\lambda K^{\lambda}}{x^{\lambda+1}}, x \geq K \\ F(x) &= \begin{cases} 1 - (K/x)^{\lambda}, & x > 0 \\ 0, & \text{otherwise} \end{cases} \end{aligned} \tag{2.24}$$

where λ is the scale parameter and K is the shape parameter, $\lambda > 0$, $K > 0$.

2. Two-state Markov chain:
A two-state Markov chain model is utilized to model the primary user traffic behavior. Here P_I and P_B represent channel transition probability to being available and busy respectively. $1 - P_I$ and $1 - P_B$ represent transition probability from state of availability to unavailability and vice versa.

The steady state probability of channel i to be busy by sending primary user's traffic is given by:

$$\Pi_B(i) = p_B(1 - p_B)$$

The average sensing time of a SU W_i is given by:

$$W_i = \sum_{L=1}^{T_{Threshold}} LP(T_s = L) \tag{2.25}$$

where, $T_{Threshold}$ is the number of slot of secondary user's maximum channel sensing tolerance and T_s is the number of slots spend by secondary user for sensing the available channel.

A control parameter C_i is used to increase or decrease DBP index according to the channel conditions and the channel bandwidth ratio in reference to other channels' bandwidth. It is also used to track the difference between the data rates of different channels. C_i value starts from one for all channels and updates as primary user appears on channel i. The successful transmission probability P_i is calculated as the percentage of successful transmitted slots to the total number of transmitted slots on channel i. The short term updates of data rate (Ri) of channel i can be given as:

$$R_i = \begin{cases} R_i(1-\alpha), & \text{if } P_i \leq 0.9; \\ R_i(1+\alpha), & \text{if } P_i > 0.9; \end{cases}$$

where i varies from 1 to N. α is the rate smoothing parameter and its value is 0.001. R_i increases with increase in successful transmission probability $P_i > 0.9$. Similarly, R_i decreases for transmission probability $P_i \leq 0.9$ which indicates bad channel condition. The long term updating of C_i is done every 50 slots according to the following rule:

$$C_i = \begin{cases} C_i - \Delta C, & \text{if} \left[\frac{R_i}{R^*} - \frac{1}{N} \sum_{j=1}^{N} \frac{R_j}{R^*} < -\varepsilon \right] \\ \\ C_i + \Delta C, & \text{if} \left[\frac{R_i}{R^*} - \frac{1}{N} \sum_{j=1}^{N} \frac{R_j}{R^*} > \varepsilon \right] \end{cases}$$

where R^* is the target rate, i is the current channel, j is the candidate channel, ε is threshold value, and ΔC is a fixed step parameter.

If T_{max} is the maximum delay allowed by secondary user and T_{avg} is the predetermined average time, then for i varying from 1 to N, the priority index η_i can be expressed as:

$$\eta_i = \begin{cases} (T_{max} - D_i)B_i, & D_i \leq T_{avg} \\ C_i(T_{max} - D_i)B_i, & \text{otherwise} \end{cases}$$

η_i represents the maximum capacity of channel i. This priority index represents the delay bandwidth product based scheme, therefore priority index increases as the DBP increases.

The channel selection when primary user appears on the current selected channel is given as:

$$Channel_i = \max\{\eta_i\}$$

This selected $Channel_i$ becomes the candidate channel k.

The performance of this DBP-based scheme is calculated to determine whether it meets the required service and reliability objectives. The effective data rate R_{eff} can be calculate from Eq. 2.26.

$$R_{eff} = \frac{\sum_{i=1}^{N} ts_i B_i}{t_{total}} \tag{2.26}$$

where ts_i is the successful transmission slot, B_i is the average bandwidth of channel i, while t_{total} is the total transmission time and can be calculated from Eq. 2.27

$$t_{total} = \sum_{i=1}^{N} ts_i + \sum_{i=1}^{N} D_i \tag{2.27}$$

where t_{si} is successful transmission slot and D_i is total delay time of secondary user i.

The total effective data rate will be maximized as long as the cognitive radio node stays over the channel with the highest DBP index.

2.2 Media Access Control

As discussed in Sect. 1.7, a media access scheme (MAC) is required among multiple secondary users who are sharing a channel so as to mitigate *access collision*. Media access control protocols can be broadly classified into two categories: *random* and *time-slotted* media access schemes, each having its pros and cons. Recently a new class of access scheme, known as *hybrid* schemes has been developed with better performance than that of the random and time-slotted access schemes.

2.2.1 QoS Aware Media Access Schemes

A QoS aware MAC protocol for cognitive radio with real time delay sensitive multimedia applications is discussed in [7]. The discussed scheme determine the set of channels for sensing and transmission of data which will satisfy the QoS requirements of the cognitive radio user. The authors also implements a priority based spectrum access scheme at the secondary user level to provide the QoS guarantee.

The protocol is utilized in a single hop cognitive radio network environment where all the secondary users can communicate with each other. The cognitive radio networks comprises of Nc data channel and 1 control channel, which is selected in the unlicensed UWB (ultra wide band) spectrum. A control channel is used to to synchronise and agree on a data channel for transmission and then tune to that data channel for communication. Each secondary user is equipped with a single transceiver and therefore can only sense one channel at a time. Each channel is modelled as an ON/OFF model. Secondary users keeps track of the slowly varying characteristics and usage patterns of the primary users.

As discussed in Sect. 1.5.1, the secondary user senses the channel for an arbitrary time, the arbitrary sensing period (ASP) which includes a basic sensing time to ensure sensing accuracy and a random slot selected from window $[0, SW_i]$. This is done to mitigate access collisions among all secondary users who are accessing the same channel. If a primary users shows up on the same channel utilized by the cognitive radio during the data transmission phase, the transmission fails and the secondary user has to vacate the channel within a given time threshold. If the secondary user wants to continue transmission of data, the secondary user has to move to some other channel for data communication.

The transmission time of the current frame of user i is given as t_i, then the probability that the current frame is transmitted over the channel k is

$$P_i^k = \frac{\beta_k}{\beta_k + \alpha_k} F_k(t_{AS_i} + t_i) \tag{2.28}$$

where $1/\alpha_k$ and $1/\beta_k$ are the mean idle and busy periods of channel k respectively, t_{AS_i} is the arbitrary sensing period selected by user i, t_i is the transmission time for the current frame, and $F_k(x)$ is the cumulative distribution function of the usage of channel k by the primary users. It is assumed that no more than one primary user will transmit at the same time on the given channel. Therefore, the transmission by a cognitive radio user is successful if no primary user appears during the $t_{AS_i} + t_i$ interval.

Once the transmission probabilities are computed, a channel sensing sequence can be designed. Two different channel sensing policies have been discussed: *greedy* and *ascending*. In the greedy technique, all the channels are sorted in the descending order and the secondary users selects channel in that order. This guarantees maximum probability of successful transmission for each secondary user which will result in increased selection of this channel by secondary users. This increase in selection will ultimately lead to increase in contention which will significantly reduce the throughput. The ascending approach allows different secondary users to select various channels in the order of the available channel probability. For a given available channel probability of channel k for user i (P_i^k), the expected transmission time over channel k is given as $E[T_i^k] = \frac{t_{AS_i}+t_i}{P_i^k}$. The secondary user selects the channels which will satisfy the condition $E[T_i^k] < \tau_i$, where τ_i is the maximum tolerable one hop delay which a secondary user can guarantee for QoS provisioning. Because of this, the delay performance can be guaranteed by the ascending policy.

To improve the QoS, service differentiation is applied in terms of size of sensing window. A cognitive radio user differentiate the traffic based on different sensing window size. A higher priority application with higher QoS constraints will have a shorter sensing period so that they have an higher chance of accessing the data channel when the opportunity appears. (i.e., $SW_{voice} < SW_{video} < SW_{data}$). Multiple levels of QoS provisioning can be defined by defining different sensing window sizes for different traffic types as shown in Table 2.3.

For a secondary user, the transmission fails for the current time slot on channel k if any of the following conditions are true:

(i) The channel is already occupied by a primary user. The probability of which can be given as:

$$P_{occupied}^k = \frac{\alpha_k}{\alpha_k + \beta_k} \tag{2.29}$$

Table 2.3 Sensing windows for different multimedia services

Type	Strict priority	Statistical priority	No priority
Voice	[0, 31]	[0, 31]	[0, 31]
Video	[32, 63]	[0, 63]	[0, 31]
Data	[64, 127]	[0, 127]	[0, 31]

(ii) A primary user occupies the channel in middle of channel sensing. The probability of which can be given as:

$$P_{PU}^{k} = \frac{\beta_k}{\alpha_k + \beta_k} \cdot F_k(t_{AS_i}) \tag{2.30}$$

(iii) The channel gets busy due to another secondary user transmission. This will depend upon the sensing window size. Here for a maximum sensing window size of W and N_k secondary users contending on channel k, the probability that the given secondary user wins the contention is given as:

$$P_r(\text{given secondary user wins contention}) = \sum_{j=1}^{W} \frac{1}{W}\left(\frac{W-j}{W}\right)^{N_k} \tag{2.31}$$

Since every secondary user is contending on the channel k, it implies that there were no primary users' detected on that channel. Therefore the probability that a secondary user's transmission fails due to another secondary user transmission is given as:

$$P_{SU}^{k} = \frac{\beta_k}{\alpha_k + \beta_k}\left(1 - F_k(t_{AS_i})\right)\left(1 - \sum_{j=1}^{W} \frac{1}{W}\left(\frac{W-j}{W}\right)^{N_k}\right) \tag{2.32}$$

Considering all three cases, the probability that the transmission from a secondary user is successful on channel k is given as:

$$P_{success}^{k} = 1 - \left(P_{occupied}^{k} + P_{PU}^{k} + P_{SU}^{k}\right) \tag{2.33}$$

With this, the average time a secondary user occupies a channel k is given as:

$$E[T^k] = P_{success}^{k}(t_{AS_i} + t_i) + (1 - P_{success}^{k})t_{AS_i} \tag{2.34}$$

A transmission is successful only if no primary user shows up during the total sensing and transmission time of the secondary user. The probability of which can be given as:

$$P^k_{TS} = \frac{\beta_k}{\alpha_k + \beta_k} \sum_{j=1}^{W} \frac{1}{W} \left(\frac{W-j}{W} \right)^{N_k} F_k(t_{AS_i} + t_i) \tag{2.35}$$

The probability that a secondary user is able to succeed in transmission in r-th attempt on the channel k is given as:

$$P_{TS}(r) = P^k_{TS} \prod_{j=1}^{r-1} (1 - P^k_{TS})^j \tag{2.36}$$

The average transmission delay of a secondary user over a given channel k is given as:

$$E[T] = \sum_{r=1}^{\infty} E[T^k] P_{TS}(r) \tag{2.37}$$

The discussed media access scheme along with channel sensing scheduling scheme is QoS aware, which provides differentiated services for multimedia and voice. It provides real time service guarantees for traffic with higher priorities. Since it is utilized for single hop network, in multiple hop network, the protocol will suffer from the problem of hidden terminal.

In [8], two media access scheme are discussed for cognitive radio networks which provisions certain level of QoS to voice packets in the network. The two media access scheme are based on *contention* and *time-division.*

The *contention* based medium access protocol is similar to various protocols that have been already discussed. The channel time is partitioned into time slots, and a time slot is further divided into four phases: *sensing, contention, transmission* and *ACK*. The first phase (sensing phase), is used for all the secondary users to sense the activity of the primary users. If the channel is sensed busy, no secondary user contends for that slot. The second phase (contention phase), comprises of a number of minislots. All the secondary users have the same contention window size. If the channel is sensed idle in the sensing phase, a secondary user randomly chooses a backoff timer from the contention window and continues to sense the channel in the contention phase. If the channel is sensed to be continuously idle for a duration of the backoff timer (in units of minislots) in the contention phase, the secondary user will terminate during the contention phase and transmit its packet in the third phase of the time slot (transmission phase); otherwise, it will quit the contention for the current slot. Therefore, for each contention, the secondary user with the smallest backoff timer will win and transmit its packet in the transmission phase of the time slot. If more than one secondary user choose the same smallest backoff timer, it will result in a collision. To determine whether a packet has successfully been transmitted, the receiver sends an acknowledgment (ACK) at the ACK phase of each slot to the sender upon a successful packet reception.

The *time-division* based media access depends on dividing the time into indexed mini slots. Each mini slot is assigned to a secondary user in a deterministic way such that the secondary user can only transmit if the channel is idle from minislot 0 to $i-1$. Therefore a secondary user has to sense the channel from 0 to $i-1$ and can only transmit if the channel is found idle. This scheme also means that the CR users with higher indexed mini slots will have a lower chance of getting a transmission opportunity and hence the indexing is rotated periodically. If the number of users in the CR network are always same, mini slot assignment can be done at initialization. If instead the number of users change dynamically, the protocol can utilize control packets for joining and leaving the network.

A 3-dimensional Markov chain is utilized to do numerical analysis where each state of the Markov chain is represented by (n, t, s), where $s = 0$ is the idle state and $s = 1$ is the busy state. When $n > 0$, t is the queuing delay experience by the queued packet, and when $n = 0$, t is the time required for the arrival of the next voice packet where n indicates the number of packets in the queue.

The probability of success for the contention based protocol in independent channel-state model is derived as:

$$P_s = \theta^{N_p} \sum_{i=0}^{N_s-1} \binom{N_s-1}{i} \rho^i (1-\rho)^{N_s-i-1} \left[\sum_{j=1}^{CW} \frac{1}{CW} \left(\frac{CW-j}{CW} \right)^i \right] \tag{2.38}$$

where θ^{N_p} is the probability that a channel is vacated by individual primary users and ρ is the queue utilization of the secondary user. N_s, is the total number of secondary users and CW is the contention window size of secondary users.

The probability of success for the contention-free (time-division) protocol in independent channel-state model is derived as:

$$P_s = \theta^{N_p} \sum_{i=1}^{N_s} \frac{1}{N_s} (1-\rho)^{i-1} \tag{2.39}$$

The packet drop probability in independent channel-state model is calculated as:

$$P_{drop} = \frac{\sum_{n>0} \pi(n,t)(1-P_s)^{D-t}}{\sum_{n>0} \pi(n,t)} \tag{2.40}$$

where $\pi(n, t)$ is the steady state probability of state (n, t). Then, the capacity for the voice secondary users N_s^* can be obtained by finding the maximum integer value of N_s that satisfies $P_{drop} \leq P_l$.

The method ensures the packet loss if the number of users are within the allowed limits along with the provisioning of certain level of QoS the the voice users. The protocol has very high overhead due to maintaining indices and dynamically changing indices in the time-division based media access scheme. The performance of the network would suffer when nodes leave and rejoins the networks.

In [9], a media access control scheme for multi-hop cognitive radio network is discussed. This MAC scheme aims at increasing the throughput in a distributed cognitive radio environment while utilizing the power saving mode operation for conserving energy.

The media access control utilizes CSMA/CA based channel access with distribution coordination function for distributed multi-channel access. The media access scheme also utilizes a power saving mode (PSM) operation for conserving power. The idea of PSM is to let the nodes enter a low powered dozing state if no communication is required. As discussed in Sect. 1.7.1, time in the system is divided into superframes, where each superframe is divided into $ATIM$ window and $Data$ window. The $ATIM$ window also comprises of a beacon interval. At the start of each beacon interval, any node A which has to send a packet to a node B sends an ad-hoc traffic indication message ($ATIM$) to node B within this window. When B receives the packet, it sends an ATIM-ACK frame and A and B stay awake for the rest of communication duration. Nodes that did not send or receive a $ATIM$ frame enter a doze state till the next beacon period. The $ATIM$ window can be represented as contention window as discussed in Sect. 1.7.1 and shown in Fig. 1.9. Each of the cognitive radio user utilizes a dedicated sensing hardware for sensing purposes. Therefore a cognitive radio user has the ability to perform *out-of-band* sensing. We will discuss about *out-of-band* sensing later in Sect. 3.4.2.

The system utilizes distributed collaborative sensing to detect primary user interference and to avoid the hidden node problem. The system comprises of N secondary users and utilizes $C + 1$ non-overlapping homogeneous channels, each with same bandwidth. A dedicated channel is utilized as common control channel for exchange of control signals. The size of the beacon-period is smaller than the size of the active period of primary users. A cognitive radio node can only perform *out-of-band* sensing on other channels.

Each cognitive radio user maintains the spectral image of the primary user in a vector defined by SIP. It also maintains a secondary channel load in a vector defined by SCL. The SIP provides the local view of the spectrum and has following meaning

(i) ($SIP[c] = 0$) indicates there is no primary user active on channel c.
(ii) ($SIP[c] = 1$) indicates that there is a primary user active on channel c.
(iii) ($SIP[c] = 2$) indicates that the activity of primary user on channel c is uncertain.

The SIP vector is used to determine if the network can user a certain channel for data communication. The SCL vector is used for choosing the communication channel, it contains the expected load of the cognitive radio communication. The spectral opportunity with the lowest SCL is picked when a node wishes to communicate.

During the $ATIM$ window, the nodes participate in the frame synchronization algorithm (for time synchronization), performing channel sensing, retrieving network wide spectral opportunities, and performing two-way control handshake for data exchange. The fast sensing is performed during the $ATIM$ window while $fast$

sensing is performed during the $Data$ window. The $fast$ sensing is used to update the SIP vector. If the SIP value of a channel is set to 2, that channel will be $fine$ sensed. During the $ATIM$ window, channel sensing results are also exchanged via scan result packets (SRP).

In a multi-hop scenario, certain operations are additionally required for the correct functionality of the cognitive radio system.

(i) A channel reservation is done during the $ATIM$ window for utilization during $Data$ window via a 3-way handshake. In the $ATIM$ frame, the sender inserts its list of channel opportunities. Receiver compares this with his own list and selects the channel with the least SCL values. The selected channels for data communication is inserted into the $ATIM - ACK$ frame. To mitigate hidden terminal problem, the transmitter rebroadcast the selected channel in the $ATIM - RES$ frame.

(ii) To mitigate the problem of out-of-order frame delivery of $ATIM$, SRP and beacon, a distributed coordination function is utilized. Different backoff windows are applied for different type of traffic. Here traffic is divided into two classes, (high priority and low priority). High-priority packets can select a backoff timer $w \in [0, CW/2)$, and low- priority packets have to select a backoff timer $w \in [CW/2, CW)$. The value of backoff timer is reset each time a packet is transmitted successfully. The beacon and SRP utilizes the high priority backoff timer while the $ATIM$ packet utilizes low priority backoff timer.

(iii) In a multi-hop scenario, the presence of a primary user should be detected by all cognitive radio user even if they are operating on different channels. If a secondary user sees the channel to be free during the current superframe and it appeared to be busy during the last superframe then a deep channel sensing is performed to check whether a channel is really free. This provides additional insurance over primary users mitigation to interference. The time required for vacating a channel is given as channel vacate time (CVT).

The miss detection probability of a secondary user is given by $p_{md}^{(i)}$. The false alarm probability of a secondary user is given by $p_{fa}^{(i)}$. The number of cognitive radio users within the detectability range of the primary user is given as M. The probability of randomly selecting a channel is given as $1/C$. The network wide miss detection and false alarm probabilities, $p_{md}^{(n)}$ and $p_{fa}^{(n)}$ can be calculated as.

$$p_{md}^{(n)} = \left(1 - \frac{1}{C}(1 - p_{md}^{(i)})\right)^{M} \tag{2.41}$$

$$p_{fa}^{(n)} = 1 - (1 - p_{fa}^{(i)})^{N/C} \tag{2.42}$$

A channel is only vacated if a cognitive radio detects at least one primary user on the channel i.e. a primary user appears inside the detectability range of the cognitive radio performing the channel sensing. The mean value of the CVT can calculated as:

$$\begin{aligned} E[CVT] &= \frac{t_{BI}}{2} + \sum_{k=1}^{\infty} (p_{md}^{(n)})^k t_{BI}[s] \\ &= \frac{t_{BI}}{1 - p_{md}^{(n)}} - \frac{t_{BI}}{2}[s] \end{aligned} \tag{2.43}$$

where t_{BI} is the length of the beacon interval.

A channel opening time is the time it takes to open up a channel on which a primary user has ceased its operation. To simplify the analysis, it is assumed that the false detection probability of cognitive radio is zero. If $s^{(i)}$ is the number of scans a single cognitive radio requires to detect a channel as free, then it can be calculated as:

$$p(s^{(i)} = x) = \frac{1 - p_{fa}^{(n)}}{C} \left(\frac{C-1}{C} + \frac{p_{fa}^{(n)}}{C} \right)^{x-1}, x \in \mathbb{N}_0^+ \tag{2.44}$$

Since, the system is utilizing a distributed sensing scheme, all cognitive radio need to sense the channel as free before it is opened for communication. The number of scans the entire cognitive radio network needs to detect the channel as free is denoted by $s^{(n)}$. Therefore,

$$\begin{aligned} p(s^{(i)} \le x) &= \sum_{k=1}^{x-1} p(s^{(i)} = k), x \in \mathbb{N}_0^+ \\ p(s^{(n)} = x) &= \sum_{m=1}^{M} \left[\binom{M}{m} p(s^{(i)} = x)^m * p(s^{(i)} \le x)^{M-m} \right], x \in \mathbb{N}_0^+ \end{aligned} \tag{2.45}$$

Hence the average channel occupancy time can be calculated as:

$$E[COT] = t_{BI} \left(\sum_{k=1}^{\infty} x p(s = x) \right) + \frac{t_{BI}}{2}[s] \tag{2.46}$$

The primary user activity is modelled as Poisson process where the ON and OFF periods are exponentially distributed with parameters μ_{ON} and μ_{OFF} respectively. The probability that a channel is free is calculated as:

$$p_{free} = \frac{\frac{1}{\mu_{open}^*}}{\frac{1}{\mu_{open}^*} + \frac{1}{\mu_{closed}^*}} \tag{2.47}$$

where μ_{open}^* and μ_{closed}^* are calculated as:

$$\mu^*_{open} = \left(\frac{1}{\mu_{ON}} + E[COT] - E[CVT]\right)^{-1}$$
$$\mu^*_{closed} = \left(\frac{1}{\mu_{OFF}} + E[COT] - E[CVT]\right)^{-1} \tag{2.48}$$

If all primary users are assumed to be inside the detection range of secondary users, then the throughput of a single hop cognitive radio network can be calculated as:

$$S_t = \sum_{i=1}^{C+1}\Bigg[{}^{C}C_{i-1}\left[p_{free}\left(1 - p^n_{fa}\right)\right] * \left[1 - p_{free}\left(1 - p^n_{fa}\right)\right]^{C+1-i}\left(\frac{t_{BI} - t_{ATIM}}{t_{BI}}\right) * \left(\left[L - mod(L,i)\right]S_c(\lfloor L\rfloor i) + mod(L,i)S_c(\lfloor Li\rfloor i + 1)\right)\Bigg] \tag{2.49}$$

where $L = N/2$, denotes the number of communication links, $\lfloor x\rfloor$ is the floor function, $mod(x, y)$ is the modulo operator, $S_c(x)$ is the throughput estimation of the given model with x as the number of links on the channel.

Increasing the number of cognitive radio nodes increases the false alarm probability and therefore there is a throughput loss. This can be compensate by increasing the number of channels. Increasing the number of the cognitive radio nodes in the vicinity of the primary users decreases the missed detection probability. This MAC scheme utilizes distributed sensing scheme and solve the problem of hidden terminal problem while providing the feature of broadcast and multicast. Since, this media access scheme is based on dedicated control channel, it suffers from control channel saturation problem. While contending over control channel the bandwidth is wasted on data channels.

2.2.2 *High Throughput Media Access Schemes*

A cognitive MAC protocol (C-MAC) for distributed multi-channel is discussed in [10], which utilizes distributed quiet periods for effective spectrum sensing. The multi-channel is synchronized by utilizing varying beacon periods in different channels. It supports group communication like broadcast and multicast packet dissipation through a common channel.

The protocol uses the concept of a rendezvous channel (RC) for coordination between the nodes of the network. All the nodes agree on using exactly one channel as RC when joining the network. Each channel time is divided into time-slots called as superframes, which in turn has a beacon period and a data transfer period. The beacon period across all channels are non-overlapping, which allows for efficient utilization of the channel. Also, this means that if any node wants to get information regarding other in all channels, the node has to switch channels taken in ascending order of beacon period start time (BPST). On the other hand, for getting information about a node and the channel it is located on, switching to the RC channel is adequate as all the nodes periodically transmit their information beacons on the RC.

The RC is utilized to manage the entire cognitive radio network in the following ways:

(i) Network wide group communication via broadcast and multicast packet transmission.
(ii) Cognitive radio nodes visit the RC regularly to keep their clocks synchronized so as they can adjust the channel beacon period start time. Also, this helps the cognitive nodes to keep track of the offset they required to shift their BPST when moving to other channels.
(iii) Each cognitive node visit the RC periodically and therefore can figure out the other nodes residing on other channels. This helps to dissipate the neighbor node information among all cognitive radio users.
(iv) During the exchange of beacons over RC, the number of beacons per channel can be utilized to figure out the load on that channel. Here load refers to the number of cognitive nodes residing on a channel.
(v) The RC also acts as entry point to any new cognitive radio node joining the network. Any new cognitive radio node which wants to join the network first searches the RC (with appropriate beacon header) and switches to it.

To join a network, a new node sends its beacon onto the RC letting other nodes know about its existence. After learning the network wide information from the RC, the new node switches to the appropriate channel and becomes part of that channel.

If no RC channel is found, the new node selects itself to make a decision on marking a channel as RC. Once the node makes a decision on particular channel as RC, it starts transmitting RC beacon on it, so that other cognitive users can join it. It is possible that two or more cognitive user starts the process of starting RC on some channels. In such case, convergence of RC is required. Cognitive nodes in such system do *out-of-band* measurements of the spectrum. This is done to detect the presence of other RCs and collecting channel quality information. If some other overlapping RC is detected then the detector cognitive node moves back to its original RC and transmit a RC-switch information in its beacon. This RC-switch information informs other users on that RC to switch to the new RC along with the designated time of moving. Other data channels are formed in a similar way by the cognitive nodes, except that they send an additional beacon information over RC regarding the channel switch information.

To dissipate the complete network information and mitigate hidden terminal problem, a form of distributed beaconing is employed in each channel by its occupants. At the beacon period of each frame, each cognitive node in its allocated slot transmits a beacon which contains information about its neighbours and neighbours of neighbours (information received in the past frames). This allows all nodes to keep track of other cognitive nodes on the same channel and allows mobility of cognitive nodes on the same channel.

All the nodes periodically (and in a non overlapping fashion) switch to RC and transmit their beacon for synchronisation purposes. This means that the information contained in the RC is almost always up to date. This also helps in channel selection. Any node in any of the channel can analyze the load on the channel from the number of node beacons transmitted in the channel. This is rebroadcasted at the RC when this node switches to RC during its resynchronization phase. Hence, other cognitive nodes comes to know about the loads on other channels too. This allows selection of appropriate channels by the cognitive nodes so as to decrease the contention and increase the throughput.

Group communication like broadcast and multicast is tackled in the protocol using the RC channel. Whenever a node has to perform a broadcast/multicast, it shifts to RC and send a beacon scheduling a group communication. The group communication is scheduled such that there is sufficient time in between for all other nodes to receive the information from RC during their resynchronisation phase and be able to schedule for group communication accordingly.

Another important aspect of this protocol is that it provide coexistence of secondary users with primary incumbents as well as other secondary users. The detection of primary users is done during quiet period (QP) defined for all channels (and non-overlapping for each). Once a primary user is detected, a beacon is transmitted to notify other on the channel about the incumbent and channel evacuation is facilitated. Coexistence with other secondary users (self-coexistence) is achieved via a beacon period merging scheme. If an overlapping foreign BP is detected within the BPST of the current cognitive node, the current cognitive node changes its BPST to the foreign BPST and adjust its beacon slot accordingly. If a non-overlapping foreign BP is detected, and its BPST lies within the first half of the current node's superframe, then the current node relocates its beacon to the foreign BP.

The scheme provides a robust coexistence mechanisms when incumbent nodes are present in the channel. It provides proper mitigation of hidden node problem, synchronization achieved in time, space and frequency via the beaconing approach. The utilization of RC provide multicast and broadcast facility.

The scheme is dependent on RC for coordination making it a bottleneck. The overhead for synchronisation and coordination becomes significant when data packets are small. Especially since a 3 way handshake is required in the data channel and for a beacon period synchronisation in the beacon period of the frame. Due to the requirement that the beacon period is non-overlapping, the frame length is a function of the number of nodes on a channel and the beacon length. This may cause a significant degradation in throughput if the packets are small compared to the transmitting window of the frame. The hidden terminal problem is addressed using neighbour and next neighbour information included in the beacons transmitted. Mobility may still cause nodes to go out of this extended awareness radius causing hidden node problems.

2.2.3 Self-coexistence Based MAC Protocol

If cognitive radio based wireless devices make a breakthrough in practical day to day life, then the opportunistic spectrum access will become more challenging. The increase in complexity will be attributed to coexistence of multiple cognitive radio networks in the vicinity of each other. Since one cognitive radio network will normally see other cognitive radio network as a foreign network or maybe as primary incumbent of the channel on which it is residing. Most of the existing MAC protocols in the literature takes a conservative approach in terms of spectrum sensing, i.e. when a spectrum is sensed as busy, the cognitive radio node ceases to transmit. In [11], a media access scheme is designed which considers the coexistence of multiple cognitive radio networks and ensures fair access of resources among all users.

The problem with such an approach is that even if the channels are occupied by other CR users from different networks, the node sensing the channel will still label it as busy and will simply exclude it. Instead of utilizing a two state model (idle or busy), a three-state sensing model for a channel is discussed.

In a system comprising of one primary user, M infrastructure based cognitive radio networks (denoted as $CR_1, CR_2, \ldots CR_M$) coexists. Each CR_i network comprises of N_i secondary users. The cognitive radio node utilizes two transceivers: one for communication and one is dedicated for spectrum sensing. Cooperative sensing is done by nodes of the CR_i network and a decision is made regarding the channel availability by the base station (BS) of the CR_i network. Let DT be the time required for cooperative spectrum sensing, therefore the channel informations are updated periodically every DT period of time. The channel sensed by the CR user (r_i) in a the three state model is given as:

$$r_i = \begin{cases} n_i & H_0 \\ x_p + n_i & H_1 \\ x_s + n_i & H_2 \end{cases}$$

where H_0, H_1 and H_2 are the state of channel being empty, channel occupied by a primary user and channel occupied by some secondary user, respectively. $x + p$ and x_s are the signals from the primary and secondary users, while n_i is the zero mean additive white Gaussian noise.

If a channel is busy, the cognitive node performs a fine sensing and analyze the received energy in the second stage via distance estimation technique to identify whether the channel is occupied by a primary user H_1 or by some other cognitive radio user H_2. When a cognitive radio user senses state H_2, it competes for the channel in a fair manner among all remaining cognitive radio nodes as ensured by the protocol. If instead it senses state H_1, it immediately vacates the channel to not cause interference to the primary user transmission.

Here primary users have strict higher priority over secondary users. If a primary user transmission is detected (H_1 state), a secondary user immediately tries to vacate the channel. This will require a certain duration of time, which appears as interference time to the primary user. There is a bound on maximum interference time that can appear to a primary user (T_{max}). If there is no primary user present on the channel (H_0 or H_2), the secondary users continue accessing the channel to determine the exact channel state (i.e. whether H_0 or H_2). If the channel state is determined as H_0, the secondary user access the channels like any other cognitive radio node. If the channel is determined in the H_2 state, then the conventional IEEE 802.11 [12] distributed coordination function (DCF) mechanism is utilized for contention and channel access as discussed in Sect. 1.5.1. If a secondary user has a packet for transmission, it waits for a DCF interframe space (DIFS) duration before transmitting the packet. If during this waiting period any other user is detected on the channel then a backoff mechanism is utilized. The backoff duration is randomly selected from $[0, W-1]$, where W is the size of contention window. To achieve the optimal fairness among all secondary users, this method does not utilize binary exponential backoff mechanism of IEEE 802.11.

If X represent a random variable denoting the time period during which a primary user continuously transmits on a channel, its probability density function is represented as $f_X(x)$. Therefore the duration during which the channel is not accessible (Y) by any secondary user is given as:

$$\begin{aligned} Y &= \left(\lceil \frac{X}{DT} + 1 \rceil\right) DT \\ &\approx X + DT \end{aligned} \tag{2.50}$$

The cumulative distribution of Y can be calculated as:

$$\begin{aligned} F_Y(y) &= P_r\big((x + DT) \leq y\big) \\ &= \int_0^{y-DT} f_X(x)dx \end{aligned} \tag{2.51}$$

Therefore, the mean of Y can be calculated as:

$$E(Y) = \int_{DT}^{DT+t} y F'_Y(y)dy \tag{2.52}$$

If X follows an exponential distribution with parameter μ, the mean value of Y can be calculated as:

$$\begin{aligned} E(Y) &= \int_{DT}^{DT+t} y\mu e^{-\mu(y-DT)}dy \\ &\approx \frac{1}{\mu} + DT \end{aligned} \tag{2.53}$$

For a primary user following Poisson traffic with mean arrival rate as λ, the probability that the channel is occupied during the time period t can be calculated as:

$$
\begin{aligned}
P_{occupied} &= \frac{\lambda t E(Y)}{t} \\
&= \left(\frac{1}{\mu} + DT\right)\lambda
\end{aligned}
\tag{2.54}
$$

Therefore the probability that the channel is free for secondary users to use is given as $P_{free} = 1 - P_{occupied}$.

A Markov chain model is developed to do the analysis of throughput saturation of the network. Where each state in the Markov chain represents the state of the backoff stage of the secondary user's transmission. Each time slot is considered either a period of primary user occupancy, successful secondary user transmission, idle or collision among multiple secondary user. The channel access probability P_α is calculated as:

$$
P_\alpha = 1 - \frac{W}{P_{free}(1 - P_\alpha)^N + P_{occupied} + W}
\tag{2.55}
$$

Therefore the probability that a transmission by secondary user over a given time slot would be successful is given as:

$$
P_s = P_{free} N P_\alpha \left(1 - P_\alpha\right)^{N-1}
\tag{2.56}
$$

The probability that a given time slot is idle slot is given as:

$$
P_{idle} = P_{free}(1 - P_\alpha)^N
\tag{2.57}
$$

The probability that a collision happens in a given time slot is given as:

$$
P_{cs} = P_{free}\left[1 - (1 - P_\alpha)^N - N P_\alpha.(1 - P_\alpha)^{N-1}\right]
\tag{2.58}
$$

If the average time of a successful secondary user transmission slot, the average time of collision slot caused by other secondary users and primary users are represented by T_s, T_{cs} and T_{cp}, then the throughput of secondary user is calculated as:

$$
\theta = \frac{P_s T_s}{P_{idle}\sigma + P_s T_s + P_{cs} T_{cs} + P_{occupied} T_{cp}}
\tag{2.59}
$$

where σ is the duration of an idle time slot.

The saturation throughput (θ^*) is calculated by optimizing the Eq. 2.59 as:

$$
\theta^* = max\{\theta\}
\tag{2.60}
$$

This work primarily focused on coexistence technique including self-coexistence which many other work fail to do. A kind of fairness scheme is utilized which provides fairness to all cognitive radio nodes in the network. Extending this scheme to an ad-hoc cognitive radio network would be tedious, as the system would have to make distributed decision over channel state. These channel states may appear different for different cognitive users within the same network.

A conceptual idea about an opportunistic spectrum access (OSA-MAC) [13] scheme was discussed in Sect. 1.7.1. This protocol utilizes a dedicated control channel for control packet interchange ans coordination between the nodes. Multiple opportunistic channels are utilized for data communication between secondary users which are owned by primary incumbents.

The time is divided into superframes, which in turn are divided into *ATIM* window and *transmission* intervals. During the *ATIM* window period, the secondary users exchange ATIM (announcement traffic indication message) on the control channel to indicate and agree on a channel for transmission. Once this has been done, the participating nodes switch to the agreed channel during *transmission* interval, where they compete with existing nodes using a RTS/CTS handshake and utilize the channel. A preemptive sensing is done on the data channel before any further action to make sure no primary users are present on it. If the case arises that primary users are present, then the secondary users employ backoff periods and wait for the primary users to vacate. The superframe period during second phase (where the secondary senses the channel for primary user activity) is such that none of the secondary users are allowed to transmit during this period (Quiet periods as discussed in Sect. 3.4.2).

The method utilizes two types of channel selection for data transmission: *random* and *opportunistic based*. In a random channel selection a random channel is selected with a probability of $1/L$, where L is the total number of data channels utilized in the system. In an opportunistic based channel selection, the spectrum availability probability is taken into consideration. If the number of secondary users is small as compared to the number of channels, it is better to allow the secondary users to exploit more available channels instead of selecting any random channel. For a threshold number of nodes at N_{th} and total number of users participating in communication N, if $N \leq N_{th}$ the opportunity based channel selection approach is employed or else the uniform channel selection method is employed. If $p_{i,j}$ is the probability that a channel j is available for secondary communication i and if $p_{i,k}$ is the probability that a channel k is available for secondary communication i, then the probability that a user i choses channel j is given as:

$$q_{i,j} = \frac{p_{i,j}}{\sum_{k=1}^{L} p_{i,k}} \tag{2.61}$$

The probability that k users selects channel j for data communication during transmission phase is given as:

$$C_{k,j} = {}^{N}C_{k}(q_j)^k(1-q_j)^{N-k} \tag{2.62}$$

The probability that exactly one user wins the contention among k users on channel j with maximum window size W is given as:

$$S_{k,j} = \frac{k}{W}\sum_{l=1}^{W}\left(1 - \frac{l}{W}\right)^{k-1} \tag{2.63}$$

Therefore throughput on a channel j can be calculated as:

$$T_j = \frac{Rt_{TI}}{t_{SF}}\sum_{k=1}^{N} C_{k,j}\, p_j\, S_{k,j} \tag{2.64}$$

where R, t_{TI}, t_{SF} and p_j are the transmission rate, time needed to transmit data on each channel, length of the superframe and channel available probability of channel j, respectively.

The total throughput for L channels is the summation of throughput on each channel and can be calculated as:

$$T = \sum_{j=1}^{L} T_j \tag{2.65}$$

The probability of collision of a secondary user transmission with a primary user is calculated via utilizing the probability of miss detection. The probability of collision is thus given as:

$$P_j^{(c)} = \sum_{k=1}^{N} C_{k,j}\left[1 - (1 - P_m)^k\right](1 - p_j) \tag{2.66}$$

where P_m is the probability of miss detection by the given secondary user k, p_j is the probability of channel j to be free.

This opportunistic channel selection method provides a better saturation throughput for cases where the total number of secondary is comparable to the number of channels. There is an upper bound on the number of users a network can accommodate given a minimum QoS standard and the sensing error derived from it. The control channel must be owned by the secondary service provider which would incur cost. As the number of secondary users increase, the superframe period needs to increase and since no transmission occurs on other channels during this period among the secondary, this causes underutilization of resources.

2.3 Energy Management

Users in a cognitive radio network may be powered primarily by a portable battery which generally has limited endurance between recharges. Therefore, energy management is an important issue that needs to be dealt in a cognitive radio network for greener communication, extended operable life of a user between recharges and to mitigate the effects of intermediate node failures [14]. Operating at low energy forces a user to degrade its observed QoS, and hence a graceful degradation need to be observed by such systems.

Energy management in a cognitive radio system depends upon multitude of things. This can be performed at three different steps of cognitive radio network: *transmission power control, spectrum decision* and *media access scheme*. By doing transmission power control, a user limits the transmission power of a transmitter and hence reduces the energy consumption. This reduces the SNR/SINR at the receiver side and hence the observed QoS. Spectrum allocation based energy management techniques employ selection of channels which require lower energy for transmission. A media access scheme decides how a channel is shared among multiple users in the network and thus its etiquette determines the activity of a user and hence its power consumption. Depending upon cooperation between different secondary user in the network, the sensing in the cognitive radio network can be classified into two types: *cooperative* and *non-cooperative*.

2.3.1 Cooperative Sensing Based Energy Efficient Spectrum Sensing

An energy efficient cooperative spectrum sensing scheme is discussed in [15], which optimizes the sensing time in a given superframe satisfying the constraints of sensing accuracy, throughput and latency. In a given superframe, there is a trade-off between sensing and transmission. Similarly, there is also a trade-off between waiting on a occupied channel to become free or switch onto another free channel.

If the sensing time is short then the sensing result is not accurate and hence detection of idle channel becomes impossible. Therefore carrying the transmission on current channel without accurate sensing may result in throughput loss and interference to primary incumbents. In other case, if the sensing time is too long then the transmission duration becomes small. If the channel is sensed to be busy then the secondary user can either switch to other channel or wait for the next time slot. If the secondary user waits, it will conserve energy but it will also increase the transmission delay and reduce throughput. Similarly, if it switches to another channel and performs data transmission, the the throughput is increased at the cost of increased energy consumption.

The time is divided into recurring superframes of equal length. Each superframe length is divided into sensing and transmission duration. The system comprises of M primary incumbent channels which are opportunistically utilized by the secondary users. Each secondary user performs *in-band* sensing to detect whether the current channel is idle or occupied. On the basis a decision is made on whether to stay on the current channel or move to another idle channel.

As given in Eq. 2.91, the received signal at i-th secondary user is given as r_i for m-th sample, where $s(m)$ is the primary users' transmitted signal with zero mean and σ_s^2 variance while $w(m)$ is noise which is modelled as a Gaussian process with zero mean and σ_w^2 variance ($\mathcal{N}(0, \sigma_w^2)$); while H_0 represents the hypothesis that the channel is available while H_1 represents the hypothesis that the channel is occupied by the primary incumbents.

If γ is the average SNR of primary user, then the probability of detection (P_d) and false alarm (P_f) for sensing time τ_s is given as:

$$
\begin{aligned}
P_d(\tau_s) &= Q\left(\frac{1}{\sqrt{2\gamma+1}}\left(Q^{-1}(\bar{P_f}) - \gamma\sqrt{\tau_s f_s}\right)\right) \\
P_f(\tau_s) &= Q\left(\frac{1}{\sqrt{2\gamma+1}}\left(Q^{-1}(\bar{P_d}) + \gamma\sqrt{\tau_s f_s}\right)\right)
\end{aligned}
\tag{2.67}
$$

where $Q(x)$ is the Gaussian tail probability function, while $Q^{-1}(.)$ is its inverse. f_s is the sampling frequency. $\bar{P_f}$ and $\bar{P_d}$ are target probability of false alarm and detection, respectively.

The minimum sensing time to achieve the desired target probability of false alarm and detection can be related as:

$$
\tau_s^{min} = \frac{1}{\gamma^2 f_s}\left(Q^{-1}(\bar{P_f}) - Q^{-1}(\bar{P_d})\sqrt{2\gamma+1}\right)^2 \tag{2.68}
$$

Here the minimum sensing time is related to the desired target probability of false alarm and detection. But this amount of sensing time may consume considerable energy and an optimization over it is required constrained to maximum energy consumption.

Based on the current channel sensing results, there exist three possibilities: current channel is *idle*, all channels are *busy*, and current channel is *busy* with at least one other channel is *idle*.

If the current channel is *idle*, then the secondary user makes a decision on to stay on the current channel. Switching onto another channel is inefficient and will result in increase energy consumption. If all M channels are *busy* (can be detected by *out-of-band* sensing), then no channel switching is done. If the current channel is *busy* and at least one another channel is *idle*, then a decision is made to minimize the energy consumption. If the secondary user switch to another channel, then there is increase in throughput at the cost of increased energy consumption. If the secondary user stays on the channel, then the energy consumption is minimized at the cost of throughput loss and increased latency.

If P_s is the probability that the secondary user stays on the current channel for the above discussed third case, then a joint optimization problem can be formulated with optimal sensing time (τ_s) as following:

$$\begin{aligned} &\min_{\tau_s, P_s} E(\tau_s, P_s) \\ &\text{subject to } P_d(\tau_s) \geq \bar{P}_d \\ &P_f(\tau_s) \leq \bar{P}_f \\ &W(\tau_s, P_s) \geq \mathbf{W} \\ &D(\tau_s, P_s) \geq \mathbf{D} \end{aligned} \tag{2.69}$$

where E is the total average energy required to transmit one packet, W is the average throughput, D is the average delay, $\mathbf{W}$ is the minimum average throughput and $\mathbf{D}$ is the maximum average delay.

If E_s and E_t represents energy of sensing and transmission in each time-slot respectively, then the total average energy consumption can be given as:

$$E(\tau_s, P_s) = N\tau_s E_s + NP_{sw}E_{sw} + T_{tr}E_t \tag{2.70}$$

where N is the total number of slots needed for one data packet transmission, E_{sw} is the energy cost for one channel switching, P_{sw} is the probability of switching to idle channel, T_{tr} is time required for one packet data transmission.

The average throughput W can be calculated as a function of τ_s and P_s with the time-slots of period T as:

$$W(\tau_s, P_s) = \frac{B_{st} + B_{sw}}{\tau_s + T} \tag{2.71}$$

where B_{stay} is the average number of bits that are transmitted in the duration $\tau_s + T$ when the secondary user chooses to stay on the same channel; while B_{switch} is the average number of bits that are transmitted in the duration $\tau_s + T$ when he secondary user chooses to switch to another channel.

B_{st} and B_{sw} can be given as:

$$\begin{aligned} B_{stay} &= P_{idle} B_T \\ B_{switch} &= P_{swidle} B_T (1 - P_{swbusy}) \end{aligned} \tag{2.72}$$

where P_{idle} is the probability that a channel is correctly sensed by the secondary user, P_{swidle} is the probability that a secondary user switches to an idle channel while P_{swbusy} is the probability that a secondary user switches to a busy channel.

The average delay can also be expressed as a function of τ_s and P_s with the time-slots of period T as:

$$D(\tau_s, P_s) = N\Big(\tau_s + TP_{wait}\Big) \tag{2.73}$$

where P_{wait} is the probability that the secondary user makes a decision of staying on the current channel and N is the number of frames required for completing the data transmission.

From Eq. 2.69, an optimal value of P_s can be achieved for energy minimization via setting, $W(\tau_s, P_s) = \mathbf{W}$ and $D(\tau_s, P_s) = \mathbf{D}$. Therefore the optimization problem in Eq. 2.69 is reduced from joint optimization to a single optimization variable τ_s.

The system assumes that all the channels are sensed at the same time with same accuracy which is not true and thus the probability of false detections and finding busy channels are affected. Especially in the heterogeneous environment where channel characteristics varies.

In conventional distributed sensing approach, the higher detection performance comes with a higher network energy consumption. In [16], a distributed sensing technique is discussed which reduces the network energy consumption by optimally choosing the sleeping and censoring parameters. The energy consumption is constrained and subjected to the the minimum detection performance and a maximum false alarm probability.

The system comprises of N cognitive sensor nodes and a fusion centre (FC) in a parallel distributed fusion configuration. In a parallel distributed fusion configuration, each cognitive sensor node makes a decision on the channel availability and sends the result to fusion centre. The fusion centre then fuses these local sensing data and makes a decision on the channel availability. Each cognitive radio node follows two policies: *sleeping policy* and *censoring policy*. The sleeping policy decides whether a cognitive radio node is awake or in sleep mode, while the censoring policy decides whether a sensed results is transmitted or not.

As given in Eq. 2.91, the received signal at the m-th sample for the i-th secondary user is given as r_i. Also, SNR at each cognitive radio node is considered to be same (γ). The cognitive radio node performs binary hypothesis on energy detection. H_0 represents the hypothesis that the channel is available while H_1 represents the hypothesis that the channel is occupied by the primary incumbents. $s(m)$ is the primary users' transmitted signal with zero mean and σ_s^2 variance while $w(m)$ is a Gaussian process with zero mean and σ_w^2 variance $(\mathcal{N}(0, \sigma_w^2))$. Therefore the received energy at the i-th cognitive radio node over T_0 observation samples is given as:

$$E_i = \sum_{k=1}^{T_0} r_i^2(k) \tag{2.74}$$

A form of thresholds (λ_1, λ_2) are applied to cognitive sensor nodes to enable censoring policy. The threshold bound $\lambda_1 < E_i < \lambda_2$, is called as the censoring region. Therefore by this rule, the local decisions made by the cognitive sensor nodes can be given as:

$$\text{result} = \begin{cases} \text{send 1 (implying } H_1\text{)}, & \text{if } E_i \geq \lambda_2 \\ \text{no decision}, & \text{if } \lambda_1 < E_i < \lambda_2 \\ \text{send 0 (implying } H_0\text{)}, & \text{if } E_i \leq \lambda_1 \end{cases} \tag{2.75}$$

The false alarm probability (P_f) and detection probabilities (P_d) can be given as:

$$P_f = \frac{\Gamma(T_0, \lambda_2/2)}{\Gamma(T_0)}$$
$$P_d = Q_{T_0}(\sqrt{2\gamma}, \sqrt{\lambda_2}) \tag{2.76}$$

where $\Gamma(n, x)$ is an upper incomplete gamma function given by $\Gamma(n, x) = \int_x^\infty t^{n-1}e^{-t}dt$, with $\Gamma(n, 0) = \Gamma(n)$, and $Q_u(n, x)$ is the generalized Marcum Q-function given by, $Q_u(n, x) = \frac{1}{n^{u-1}} \int_x^\infty e^{-\frac{t^2+n^2}{2}} I_{u-1}(nt)dt$. $I_{u-1}(.)$ is the Bessel function of the first kind with the order $u - 1$.

The cost function is modelled as the average energy consumed by each cognitive radio node in the system, which can be given as:

$$C_T = (1 - \mu)\sum_{i=1}^{N}(C_{s_i} + C_{t_i}(1 - \rho)) \tag{2.77}$$

where, C_{s_i} and C_{t_i} are the energy required for sensing and transmission by i-th cognitive sensor node, respectively. μ is the sleeping rate while ρ is the censoring rate and equals to the probability such that $\lambda_1 < E_i < \lambda_2$.

The constraints are applied over the global detection (Q_D) and false alarm (Q_F) probabilities. Therefore the optimization problem can be formulated as:

$$\begin{aligned} &\min_{\mu,\lambda_1,\lambda_2} C_T \\ &\text{such that } Q_D \geq \beta \\ &\qquad\qquad\; Q_F \leq \alpha \end{aligned} \tag{2.78}$$

where β and α are parameters that can be modified for different use case scenario.

Depending on the prior knowledge about the probabilities of hypotheses H_0 and H_1, two different cases are discussed.

2.3.1.1 Unknown Prior

In case of unknown prior, it can be assumed that $Pr(H_0) \lll Pr(H_1)$ (probability of hypothesis H_0 is very less than hypothesis H_1) implying that channel is mostly free from primary incumbents. The problem then can be formulated as blind Neyman–Pearson setup, which can be given as

$$\rho_{Neyman} = Pr(\lambda_1 < E_i < \lambda_2 | H_0) \tag{2.79}$$

From Eq. 2.76, the above equation can be rewritten as:

$$\rho_{Neyman} = \frac{\Gamma(T_0, \frac{\lambda_1}{2})}{\Gamma(T_0)} - \frac{\Gamma(T_0, \frac{\lambda_2}{2})}{\Gamma(T_0)} \tag{2.80}$$

From Eqs. 2.77 and 2.80, the Eq. 2.78 can be given as:

$$\min_{\mu,\lambda_1,\lambda_2} (1-\mu)\sum_{i=1}^{N}\left[C_{s_i}+C_{t_i}\left(1-\frac{\Gamma(T_0,\frac{\lambda_1}{2})}{\Gamma(T_0)}+\frac{\Gamma(T_0,\frac{\lambda_2}{2})}{\Gamma(T_0)}\right)\right]$$
$$\text{such that } Q_D \geq \beta \qquad Q_F \leq \alpha \tag{2.81}$$

The global probability of false alarm (Q_F) in the Neyman–Pearson setup can be calculated as:

$$Q_F=\sum_{k=1}^{N}\binom{N}{k}\mu^{N-k}(1-\mu)^k*\sum_{l=1}^{K}\binom{K}{l}\rho^{K-l}(1-\rho)^l[1-(1-P_f)^l] \tag{2.82}$$

Using binomial expansion theorem this can be reduced to

$$Q_F=1-\left[1-P_f(1-\mu)(1-\rho)\right]^N \tag{2.83}$$

The global probability of detection (Q_D) in the Neyman–Pearson setup can be calculated as:

$$Q_D=\sum_{k=1}^{N}\binom{N}{k}\mu^{N-k}(1-\mu)^k*\sum_{l=1}^{K}\binom{K}{l}\delta^{K-l}(1-\delta)^l[1-(1-P_d)^l] \tag{2.84}$$

Using binomial expansion theorem this can be reduced to

$$Q_D=1-\left[1-P_d(1-\mu)(1-\delta)\right]^N \tag{2.85}$$

Therefore by utilizing the value of Q_F and Q_D, the optimization problem in Eq. 2.81 can be solved to derive optimum sleeping rate (μ) and censoring thresholds (λ_1, λ_2).

2.3.1.2 Known Prior

If the probabilities of hypothesis H_0 ($Pr(H_0)$) and hypothesis H_1 ($Pr(H_1)$) are known in advance, then the problem can be formulated as knowledge aided Bayesian setup. Therefore the censoring rate can be given as:

$$\rho_{Bayesian}=\pi_0 Pr(\lambda_1<E_i<\lambda_2|H_0)+\pi_1 Pr(\lambda_1<E_i<\lambda_2|H_1) \tag{2.86}$$

where π_0 and π_1 are the probability of hypothesis H_0 and H_1, receptively.

From Eq. 2.76, we have:

$$
\begin{aligned}
Pr(\lambda_1 < E_i < \lambda_2|H_0) &= \frac{\Gamma(T_0, \frac{\lambda_1}{2})}{\Gamma(T_0)} - \frac{\Gamma(T_0, \frac{\lambda_2}{2})}{\Gamma(T_0)} \\
Pr(\lambda_1 < E_i < \lambda_2|H_1) &= Q_{T_0}(\sqrt{2\gamma}, \sqrt{\lambda_1}) - Q_{T_0}(\sqrt{2\gamma}, \sqrt{\lambda_2})
\end{aligned}
\tag{2.87}
$$

From Eqs. 2.77 and 2.87, Eq. 2.78 can be rewritten as:

$$
\begin{aligned}
&\min_{\mu,\lambda_1,\lambda_2} (1-\mu) \\
&\sum_{i=1}^{N}\left(C_{s_i} + C_{t_i}\left(1 - \pi_0\left(\frac{\Gamma(T_0, \frac{\lambda_1}{2})}{\Gamma(T_0)} - \frac{\Gamma(T_0, \frac{\lambda_2}{2})}{\Gamma(T_0)}\right)\right.\right. \\
&\left.\left. + \pi_1\left(Q_{T_0}(\sqrt{2\gamma}, \sqrt{\lambda_1}) - Q_{T_0}(\sqrt{2\gamma}, \sqrt{\lambda_2})\right)\right)\right) \text{ such that } Q_D \geq \beta \\
&Q_F \leq \alpha
\end{aligned}
\tag{2.88}
$$

The local probability of false alarm and detection in knowledge aided Bayesian setup can be given as:

$$
\begin{aligned}
P_f^B &= (1-\mu)(1-\delta_0)P_f \\
P_d^B &= (1-\mu)(1-\delta_1)P_d
\end{aligned}
\tag{2.89}
$$

Therefore, the global probability of false alarm and detection in knowledge aided Bayesian setup can be derived as:

$$
\begin{aligned}
Q_F &= 1 - \left[1 - P_f(1-\mu)(1-\delta_0)\right]^N \\
Q_D &= 1 - \left[1 - P_d(1-\mu)(1-\delta_1)\right]^N
\end{aligned}
\tag{2.90}
$$

where $\delta_0 = Pr(\lambda_1 < E_i < \lambda_2|H_0)$ and $\delta_1 = Pr(\lambda_1 < E_i < \lambda_2|H_1)$.

Therefore by utilizing the value of Q_F and Q_D, the optimization problem in Eq. 2.88 can be solved to derive optimum sleeping rate (μ) and censoring thresholds (λ_1, λ_2).

The blind Neyman–Pearson setup is shown to be an special case of Bayesian setup. As the transmission energy increases compared to the sensing energy, the total transmission energy has to be reduced more than the sensing energy thus the censoring rate increases and the sleeping rate decreases. It is shown that without sleeping or censoring, the energy consumed saturates to a level which is much lower to conventional energy consumption. Also, as the number of users increase, the optimal sleeping rate increases dramatically to keep the energy consumption of the system stable.

2.3.2 *Non-cooperative Sensing Based Energy Efficient Spectrum Sensing*

Traditionally usage of cooperative sensing provides efficient channel occupancy results in a resource constrained environment, but it requires exchange of control messages. This exchange of control messages for cooperation increases the energy consumption of cognitive radio network. In [17], a time division energy efficient (TDEE) non-cooperative spectrum sensing scheme for cognitive radio network is discussed. In this scheme, the sensing period is divided in time slots and each secondary user in the network is assigned a different channel to detect in each slot. This allows secondary users to accurately sense the network without exchanging control messages for cooperation and allowing to save considerable energy.

The system utilizes a central controller node (base station in infrastructure network or cluster head in case of ad-hoc network) which coordinates the cognitive nodes in the network. The central controller designates cognitive nodes in the network to perform sensing on given channels along with the given sensing order. The information is then received by the central controller from the cognitive nodes, which in turn perform fusion on the received spectrum sensing data to make a decision on the availability of channels. The system time is divided into superframes, where each superframe comprises of two parts: *sensing* and *transmission*. The duration of the superframe is given as T, in which T_s is the time duration of spectrum sensing whereas T_r is the time duration of transmission in the superframe. The time T_s is further divided in equal time slots during which the secondary user senses different channels in the order given by the central controller. After the expiration of sensing period all the SUs send the sensing results to the central controller. For given N secondary users and M channels in the system, the given scenario considers where U secondary users want to perform sensing, where $2 \leq U \leq min(N, M)$.

The steps required for a secondary user to perform the sensing in the discussed scheme can be given as:

(i) All secondary users join to the network controller by central controller by following their network join/leave protocol. Once joined, it waits for the control messages from the central controller.
(ii) The central controller provides the information on number of channels to be sensed by the secondary users. Depending upon that, the sensing duration is divided into mini time-slots during which the secondary user individually scans each channel. Once the sensing is performed on all channels, the secondary users will update the information to the central controller.
(iii) The central controller fuses the channel availability information received from all secondary users and based on that selects a channel.

Since, the control packets are only exchanged with central controller, the number of control packets exchanges is substantially reduced. The efficiency of the discussed can be analysed in two schemes: *Homogeneous* networks and *Heterogeneous* networks.

2.3.2.1 Homogeneous Networks

In a cognitive radio based homogeneous network environment, all secondary users have the same sensing time and same channel coefficient h from primary incumbent's transmitter. The sensing duration is divided in U time slots and length of each slot is τ such that $T_s = U\tau$.

Total energy spent in the sensing duration is sum of energy spent for sensing different channels and energy used for exchange messages. It is clear that the sensing/exchange energy will not only depend on the sensing/exchange power but also on sensing/exchange time. Although the exchange duration is very small the exchange energy can be neglected, but as the number of cooperative secondary users increase, the exchange energy will also increase (keeping in mind retransmission if error happens). Therefore the energy consumed via exchange of control packets is also taken into consideration. The received signal r_i at the m-th sample for the i-th secondary user can be given as:

$$r_i(m) = \begin{cases} w(m), & H_0 \\ s(m) + w(m), & H_1 \end{cases} \tag{2.91}$$

where H_0 represents the hypothesis that the channel is available while H_1 represents the hypothesis that the channel is occupied by the primary incumbents. $s(m)$ is the primary users' transmitted signal with zero mean and σ_s^2 variance while $w(m)$ is a Gaussian process with zero mean and σ_w^2 variance ($\mathcal{N}(0, \sigma_w^2)$).

The detection probability can be calculated by utilizing the central limit theorem over the Chi-square distribution and is given as:

$$P_i^d = Q\left(\frac{\lambda - \tau f_s(|h|^2\sigma_s^2 + \sigma_w^2)}{\sqrt{2\tau f_s}(|h|^2\sigma_s^2 + \sigma_w^2)}\right) \tag{2.92}$$

Similarly, the false alarm probability can be calculated as:

$$P_i^f = Q\left(\frac{\lambda - \tau f_s\sigma_w^2}{\sqrt{2\tau f_s}\sigma_w^2}\right) \tag{2.93}$$

where, $Q(x) = \frac{1}{\sqrt{2\pi}}\int_x^\infty e^{-\frac{t^2}{2}}dt$ is the tail probability of the standard Normal distribution function and f_s is the sampling frequency.

The detection probability and false alarm probability calculated by the central controller can be given as:

$$\begin{aligned} P^d(U) &= 1 - (1 - P^d)^U \\ P^f(U) &= 1 - (1 - P^f)^U \end{aligned} \tag{2.94}$$

If P_{tx} and P_{rx} are the transmitted power and received power at each secondary user, respectively, then the optimization problem of calculation of requirement of number of secondary user for energy minimization can be calculated as:

$$\min_{U} : U\Big[T_s P_{tx} + T_e P_{rx}\Big]$$
$$\text{such that, } P^d(U) \geq P^d_{Th} \tag{2.95}$$
$$P^f(U) \leq P^f_{Th}$$

where $T_s P_{tx}$ is the sensing energy consumed in the homogeneous network and $T_e P_{rx}$ is energy consumed by exchanging control messages in which T_e is the control message exchange time. $P_{d,Th}$ is the threshold of the cooperative detection and $P_{f,Th}$ is threshold of false alarm.

By applying bounds on the constraints of Eq. 2.95, the bounds on U can be given as:

$$\frac{ln(1 - P^d_{Th})}{ln(1 - P^d)} \leq U \leq \frac{ln(1 - P^f_{Th})}{ln(1 - P^f)} \tag{2.96}$$

Since, the optimization Eq. 2.95 is a decreasing function of the parameter U, the minimal value of U for minimal energy consumption is given as:

$$U = \frac{ln(1 - P^d_{Th})}{ln(1 - P^d)} \tag{2.97}$$

If P_s is the probability that a channel is successfully found then it can be given as:

$$P_s = p\Big[1 - P_f(U)\Big] \tag{2.98}$$

where p is the channel available probability in a 2-state Markov chain model of spectrum occupancy.

Therefore the probability that the available channels can be found in one sensing period ($P_{available}$) can be given as:

$$P_{available} = \sum_{u=1}^{U} \binom{U}{u} (1 - P_s)^{U-u} P_s^u \tag{2.99}$$

The throughput of a channel can be derived by utilizing the Eq. 2.99:

$$\mathbf{TH} = (T - U\tau) P_{available} R \tag{2.100}$$

where R is the bandwidth of the channel.

2.3.2.2 Heterogeneous Networks

In heterogeneous networks all secondary users have different sensing periods and different channel coefficients ($h_i, \forall i \in U$). In this case the total energy consumed will be dependent on the number of secondary users (U) along with the sensing time of each secondary user. If τ_i is the size of the time slot of i-th secondary user, then the optimization equation for minimizing the energy consumption can be given as

$$\begin{aligned} &\min_{U;T_s^1,\ldots T_s^U} \sum_{i=1}^{U} \left[T_s^i P_{rx} + T_e P_{tx} \right] \\ &\text{such that, } P^d(U) \geq P_{Th}^d \\ &\qquad\qquad P^f(U) \leq P_{Th}^f \end{aligned} \tag{2.101}$$

The Eq. 2.101 can be refined for the condition that $\sum_{i=1}^{U} T_s^i$ is minimized when all T_s^i are minimized and therefore can be rewritten as:

$$\begin{aligned} &\min_{U;\tau_i} T_s^i = U\tau_i \\ &such\,that,\; P^d(U) \geq P_{Th}^d \\ &\qquad\qquad P^f(U) \leq P_{Th}^f \end{aligned} \tag{2.102}$$

A bound on the value of U can be calculated as:

$$\frac{ln(1 - P_{Th}^d)}{ln(1 - P_{i,min}^d)} \leq U \leq \frac{ln(1 - P_{Th}^f)}{ln(1 - P_{i,max}^f)} \tag{2.103}$$

The optimization problem of finding the optimal sensing timeslots can be given as:

$$\begin{aligned} &arg \max_{\tau_i} \frac{Q\left(\frac{\lambda - \tau f_s(|h_{min}|^2\sigma_s^2 + \sigma_w^2)}{\sqrt{2\tau f_s}(|h_{min}|^2\sigma_s^2 + \sigma_w^2)}\right)}{\tau_i} \\ &\text{such that, } U \leq \frac{ln(1 - P_{Th}^f)}{ln(1 - P_{i,max}^f)} \end{aligned} \tag{2.104}$$

The optimal value of the sensing time slot is calculated as:

$$\tau_i^* = \frac{1}{2f_s}\left(\sqrt{\frac{\lambda^2}{4(|h_{min}\sigma_s^2 + \sigma_w^2) + 1}} - 1\right) \tag{2.105}$$

Similar to the homogeneous environment case, the throughput in heterogeneous environment can be calculated as:

$$\mathbf{TH} = \left(T - \sum_{u=1}^{U} \tau_i\right) P_{available} R \tag{2.106}$$

where, R is the bandwidth of each channel and $P_{available}$ is the probability that the available channels can be found in one sensing period.

The energy consumed by this approach is always less than the traditional cooperative scheme. The throughput of this scheme is much higher than the traditional scheme as during the same sensing duration, the number of channels discovered correctly is always more.

A distributed resource allocation based energy efficient scheme is discussed in [18], for cognitive radio based wireless sensor network operating in multi-carrier environment. The discussed scheme allows subcarrier selection and power allocation among individual cognitive radio nodes. The overall goal of the scheme it to reduce the energy consumption per bit over all subcarriers while maintaining the certain level of data transmission rate.

Multiple users are allowed to share the same subcarrier as long as their Signal to Interference and Noise Ratio (SINR) is acceptable. The problem of subcarrier (channel) allocation is converted into an unconstrained optimization problem and then branch and bound method of solving optimization equations is utilized to find an optimal solution.

As discussed in Sect. 1.7.2, the system time is divided into time slots where the size of each time slot is T_s. During each time-slot the cognitive nodes exchange beacons which is utilized for synchronization purposes. The entire frequency spectrum is divided into M subcarriers of which each spectrum experience flat Rayleigh fading.

After each user senses the network and obtains the list of available subcarriers, the scheme will select subcarriers and allocate power to them such that the energy efficiency is maximized, and data rate and power constrains are satisfied. If $P_t^{(i)}$ is the transmission power allocated to sub carrier i, then the optimization problem of power allocation in distributed subcarrier can be given as:

$$\min_{P_t^{(i)}} \frac{\sum_{i=1}^{M} P_t^{(i)} + P_r}{B.\sum_{i=1}^{M} log_2\left(1 + \alpha_i P_t^{(i)}\right)} \tag{2.107}$$

Subject to,

$$\begin{aligned} &\sum_{i=1}^{M} R^{(i)} = B.\sum_{i=1}^{M} log_2\left(1 + \alpha_i P_t^{(i)}\right) \geq R_{req}, \forall i = 1, 2, \ldots, M \\ &\sum_{i=1}^{M} P_t^{(i)} = P_{max}, P_t^{(i)} \geq 0, \forall i = 1, 2, \ldots, M \end{aligned} \tag{2.108}$$

where P_r is the total circuit power consumption, B is bandwidth of each subcarrier, P_{max} is the maximal power bound, R_{req} is the target data rate, α_i is channel state information of subcarrier i and M is the number of available channels.

The optimization Eq. 2.107 is not quasi-concave/convex and therefore to solve it, the constraints from this Equations are removed initially. Once the optimal solution to unconstrained problem is known, the initial constraints of Eq. 2.108 can be examined again.

On removing the constrains from the Eq. 2.107, the optimal transmission power $P^* = P_t^{(i)^*}, \forall i = 1, 2, \ldots M$ can be given as [18]:

$$
\begin{aligned}
P_t^{(i)^*} &= \max\left[B.log_2^e.\zeta^* - \left(\frac{1}{\alpha_i}\right), 0\right] \\
\zeta^* &= \min_{P_t^{(i)^*}} \frac{\sum_{i=1}^{M} P_t^{(i)} + P_r}{B.\sum_{i=1}^{M} log_2\left(1 + \alpha_i P_t^{(i)^*}\right)}
\end{aligned}
\tag{2.109}
$$

where ζ^* is optimal energy per bit.

Multiple solutions exist for Eq. 2.109. The constraints of Eq. 2.108 are examined. For two constraints, a total of four possibilities exists:

(i) If $\sum P_t^{(i)^*} \leq P_{max}$ and $\sum R^{(i)} \geq R_{req}$, then Eq. 2.109 is satisfied.

(ii) If $\sum P_t^{(i)^*} \geq P_{max}$ and $\sum R^{(i)} \leq R_{req}$, it implies that even after allocating all the channels which exceeds the maximal power bound, the target data rate requirement is not met. Therefore in such condition, there exist no feasible solution to the problem.

(iii) If $\sum P_t^{(i)^*} \leq P_{max}$ and $\sum R^{(i)} \leq R_{req}$, it implies that the target data rate requirement is not met. Since power allocated to subcarrier still hasn't reached the maximal bound, power should be increased till the maximal power bound is reached. The minimal required additional power to achieve the data rate of R_{req} can be given as:

$$
\Delta P_{min} = M\beta\left[e^{\frac{R_{req} - \sum_{i=1}^{M} R^{(i)^*}}{Blog_2^e M}} - 1\right] \tag{2.110}
$$

If even after this additional power, if $\sum P_t^{(i)^*} + \Delta P_{min} \geq P_{max}$, then there exist no feasible solution to the Eq. 2.107 with constraints in Eq. 2.108.

(iv) If $\sum P_t^{(i)^*} \geq P_{max}$ and $\sum R^{(i)} \geq R_{req}$, it implies that total power allocation and data rate both exceeds their upper bound. In this case, the power allocated in subcarrier is decreased so as to make it equal to or below the P_{max} value.

The optimal subcarrier selection and power allocation is done independently by all new users, which may lead to multiple users deciding to use the same sub-carrier and thus causing interference on it. Therefore a power control scheme based on distributed co-channel interference mitigation is required. If $N^{(i)}$ new users want to use the same subcarrier i at the same time, then the following distributed power control iteration will converge if a feasible solution exist.

$$P_n^{(i)}(k+1) = \frac{\gamma_n^{req,(i)}}{\gamma_n^{(i)}(k)} P_n^{(i)}(k) \tag{2.111}$$

Here $P_n(i)(k)$, $\gamma_n^{(i)}(k)$ are transmission power and measured SINR of new user n on subcarrier i in step k, respectively. $\gamma^{req,(i)}$ is the target SINR and is calculated from the optimal allocation data rate $R_n^{(i)*}$.

Each node only needs to know its received SINR at the receiver and use it to change its transmission power. The receiving node returns this as the feedback through a control channel. If the power control algorithm cannot resolve the interference issues then it should be resolved by the medium access control (MAC) scheme which needs to be developed. Each user can only detect interference from other user after they have started transmitting. Also it may take a lot of steps for the iterative algorithm to converge which can waste a lot of time.

In [19], an energy efficient source and channel sensing (JSCS) scheme is discussed for cognitive radio sensor networks. The overall work of the scheme is divided into two tasks: *ambient sensing* for vacant channels and *application sensing* for collecting application oriented information and delivering it to the access point/gateway. Both of these tasks consume ample amount of energy, therefore the objective of the discussed scheme is to reduce the total energy consumption while maintaining a bound on the distortion of the application-specific source information.

The ambient oriented channel sensing performs periodic channel sensing so as to detect the available channels for transmission. This helps to detect and determine the energy distribution of the spectrum and share the information with other cognitive devices in the network for cooperation. The application oriented source sensing involves invoking a sensor in the cognitive sensor node (sensors like temperature, humidity, luminosity etc.) and passing the obtained information to the access point/gateway.

The cognitive sensor network perform the two tasks of channel sensing and application sensing simultaneously. The objective of the system is to minimize total consumed energy in the system by doing a trade-off between the energy consumed while doing the channel sensing and while doing the application sensing. If excessive resources are given to application sensing then even after the collection of precise application information, it cannot be transmitted in due time due to the lack of channel sensing information. If channel sensing is given excessive resources then channel availability can be determined reliably but then the application related information might not be delivered at an acceptable rate to the access point. Thus there exist a trade-off between these two tasks.

The system time is divided into recurring superframes. The time period of each superframe is T and is short enough for a primary incumbent activity to remain unchanged in this timeframe. At the start of each superframe each node in the cognitive sensor node performs channel sensing via detection of N samples on the spectrum. The cognitive sensor network operates on carrier frequency f_c and bandwidth W and performs N sample of this channel at the starting of each superframe.

The received signal at any cognitive node can be expressed as shown in Eq. 2.91. Here $r(m)$ is the received energy at m-th sample, with $s(m)$ as the primary users' transmitted signal with zero mean and σ_s^2 variance while $w(m)$ as a Gaussian process with zero mean and σ_w^2 variance. Therefore the energy detector for N samples can be given as:

$$T(r) = \frac{1}{N}\sum_{m=1}^{N} |r(m)|^2 \tag{2.112}$$

For sufficient number of samples, the probability density function (T(r)) via central limit theorem can be expressed as:

$$T(r) = \begin{cases} \mathcal{N}(\mu_0, \sigma_0^2), & H_0 \\ \mathcal{N}(\mu_1, \sigma_1^2), & H_1 \end{cases} \tag{2.113}$$

As discussed in Eq. 2.91, H_0 and H_1 are hypothesis for the absence and presence of the primary user on the given channel, respectively. Here the primary user transmitted signal is modelled as MPSK complex signal.

If p_{false} and p_{detect} are the probability of false alarm and detection respectively, then the probability of missed detection is given as, $p_{miss} = 1 - p_{detect}$, and the probability that the cognitive node is allowed to transmit on a given superframe is calculated as:

$$p_t' = \left(1 - p_{false}\right)p(H_0) + \left(1 - p_{detect}\right)p(H_1) \tag{2.114}$$

where $p(H_0)$ and $p(H_1)$ are inactive and active probabilities of primary incumbents.

If collision probability is given as $p_{collision} = p_{miss}\,p(H_1)$, then the effective transmission probability for a cognitive node is given as:

$$p_t = p_t' - p_{miss}\,p(H_1) \tag{2.115}$$

If the energy consumed by the detection on each sample is denoted as E_{sample}, then the average channel sensing power consumption on a single superframe can be given as:

$$P_{ch-sensing} = \frac{E_{sample}N}{T} \tag{2.116}$$

The transmission power with respect to effective transmission probability p_t can be given as:

$$P_{ch-sensing}(p_t) = \left[\frac{Q^{-1}\left(1 - \frac{p_t}{p(H_0)}\right) - \sqrt{\frac{2\sigma_s^2}{\sigma_w^2} + 1}Q^{-1}(1 - p_{miss})}{\frac{\sigma_s^2}{\sigma_w^2}} \right]^2 \frac{E_{sample}}{T} \tag{2.117}$$

where Q^{-1} is the inverse tail probability of the standard normal distribution.

The application sensing power comprises of energy consumption in sensor activation, source-channel coding and transmission. The energy required for correct delivery of each bit can be given as:

$$E_{bit} = N_0 W \left[\frac{2^{(R_{channel}/W)} - 1}{R_{channel}} \right] \tag{2.118}$$

where $R_{channel}$, N_0 and W are maximum channel capacity, unilateral noise power spectral density and channel bandwidth, respectively.

Therefore the average application sensing power can be given as:

$$P_{app-sensing}(p_t) = p_t' E_{bit} R_{channel} \tag{2.119}$$

For a multi-terminal source coding system, the Eq. 2.119 can be rewritten as:

$$P_{app-sensing}(p_t) = N_0 W \left(p_t + p_{miss} p(H_1)\right) \left[\left(\frac{(\frac{\sigma_s^2}{D})^{1/K}}{1 - (\frac{\sigma_w^2}{K})(\frac{1}{D} - \frac{1}{\sigma_s^2})} \right)^{\frac{L}{2p_t W}} - 1 \right] \tag{2.120}$$

where K is number of nodes and D is distortion. σ_s^2 and σ_w^2 are the variance of the source and the variance of noise, respectively.

If the total power of a cognitive node is constrained then a trade-off is done between the application sensing and channel sensing. The effective transmission probability is the key that connects the two parameters. Therefore a joint transmission model is considered where total energy consumed can be given as:

$$P_{total}(p_t) = P_{app-sensing}(p_t) + P_{ch-sensing}(p_t) \tag{2.121}$$

If the probability of false alarm is kept below 0.5 then $P_{total}(p_t)$ forms a convex curve with respect to p_t and therefore an optimal solution exist for minimal total power consumption. By utilizing the Eq. 2.121, an optimization can be done over the application sensing and channel sensing power so as to satisfy a given distortion constraint of the source coding.

Table 2.4 Comparison of simulation, emulation and real-time test beds

	Simulation	Emulation	Real-time test beds
Implementation	Based on probabilistic and analytical modelling of components	Based on simulated and some real components	Based on realtime system components
Behavior	Results are macroscopic and moderate in terms of accuracy	Results are quite accurate and near-realistic	Results are realistic
Replication	Replication is improbable due to varying parameters	Replication to near value can be achieved	Replication can be achieved via utilizing the same components
Realization complexity	Minimal	Modest	High
Computation complexity	Very high	High	Minimal
Cost	Minimal	Modest	Expensive

2.4 Cognitive Radio Platforms

The performance of a cognitive radio system may be evaluated via either three method of implementations: simulation, emulation and real-time test beds. Simulation allows performance analysis of the overall system into an abstract level while providing macroscopic facts and figures. Emulation involves utilization of real-time (real-system) component integrated into a simulation system thereby providing a more accurate form of analysis. The small-scaled deployment of a real-time test-bed allows providing realistic results and accurate microscopic behavior of the overall system [20]. Each of these techniques have their pros and cons. Table 2.4 provides pros and cons of each of these techniques.

In wireless environment, simulation model normally misses the complex channel characterization. The wireless channels are complex in terms of shadowing, fading, noise,[1] interference and multi-path propagation. Collective modelling of these parameters poses deviation from realistic behavior which can only be solved via using realistic components in the evaluation system. Emulation tend to solve this problem by utilizing realistic components instead of complex models in the evaluation system. This provides more realistic results at the cost of utilization of realistic components. To provide accurate realistic results, real-world platform need to be deployed as test-beds [21, 22]. Therefore development of platforms for test-bed is important for evaluation of cognitive protocols.

[1]Here we refer noise as intrinsic noise, which is generated by the communication device itself; while we refer interference as the extrinsic noise, which the communication device receive from other unintended signal sources.

There has been various advances in realizing a real-time, at-scale validations of the cognitive radio technologies [22–24]. To realize such platform, generally a frequency agile flexible radio is preferred which is capable of generating multitude of waveforms controlled via software. This is normally achieved via a software defined radio. These platforms tend to accommodate design and development specifications that cover up the current scenario and scenarios yet to come. Most of the platforms are designed to be modular, so as to allow ease in development, debugging and multiple iterations of verification. The modular design also allows the platform to be structured and clean slate for easy learning and additional component building.

2.4.1 From FPGAs to Software Defined Radio

One of the requirements is that cognitive radio systems change its behaviour according to its context autonomously, based on experience and learning. Therefore the system should be able to interact with the reconfigurable network node. Also, since real time sensing is computationally intensive, it requires for high demanding digital signal processing algorithms.

Until recently, reconfigurable logic and Field Programmable Logic Array (FPGA) chips have been used primarily for prototyping. However, it is now being realized that reconfigurable architectures are a separate family in their own right, just as processors and Application specific Integrated Circuits (ASIC) are with unique properties that can be taken advantage of. General Purpose Processors are highly programmable platforms, however the performance is limited. The basic building block of FPGA is a reconfigurable cell which is implemented using Look-Up-Table (LUT), Programmable Logic Array (PLA) or Memory. LUT or memory-based cell is configured by writing the result of the combinational logic to the cell [25]. That is some modules need to be selectively reconfigured or reassigned for new mode of operation in software defined radio. Therefore, reconfigurable architectures will give greater flexibility in such an environment [24].

2.4.2 From Software Defined Radio to Cognitive Radio

Cognitive radio was first conceptualized from software defined radio by J. Mitola while his work on the field of cognitive radio [26]. The capabilities of SDR to tune to any frequency, selecting any band filter, generation of any modulation via software allows the realization of cognitive radio. It is well known that CR are computationally intensive systems that implement highly demanding digital signal processing algorithms on different platforms. Design of mutli-mode platforms using FPGA based SDR would be ideal for cognitive radio architectures. The possibility of such a system has been investigated in detail [27]. A comparative study of SDR based cognitive radio platform is done in [23].

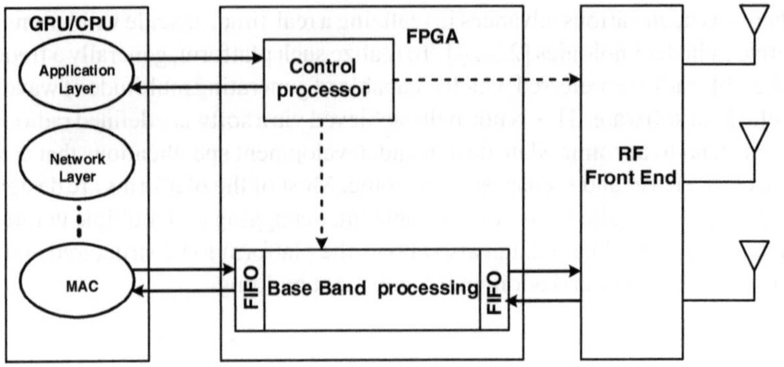

Fig. 2.3 Block schematics of SDR

The component therefore which translate SDR into Cognitive Radio is Cognition. Apart from cognition, other software components like protocol stacks are required to realize the full functioning cognitive radio network. In the next section we will discuss about different SDR control software that are utilized to emulate the cognition behavior.

Figure 2.3 shows a basic block schematic of Software defined Radio (SDR) architecture. Fundamentally, the idea of SDR is to replace most of the analog signal processing in the transceivers with digital signal processing in order to provide the advantage of flexibility through reconfiguration or reprogramming. This will enable different air-interfaces to be implemented on a single generic hardware platform. A multi-mode SDR handset with dynamic reconfigurability has the promise of integrated services and global roaming capabilities. In order to become reality, the SDR technology requires more research not only from the perspective of wireless systems, but also in the areas of novel computing architectures, embedded systems, and design methodologies to realize an all-digital reprogrammable radio.

2.4.3 SDR Software

Software controls are required for providing control over the SDR hardware. These software controls can be utilized to program a SDR to behave like a radio with cognition. Some of these SDR software are: GNU Radio Software, Pothosware, IRIS, OSSIE [23].

Figure 2.4 shows the block diagram of an SDR with an application software controlling it. The application software controls the radio behavior including frequency selection, modulation, bandwidth, gain, noise filtering etc.

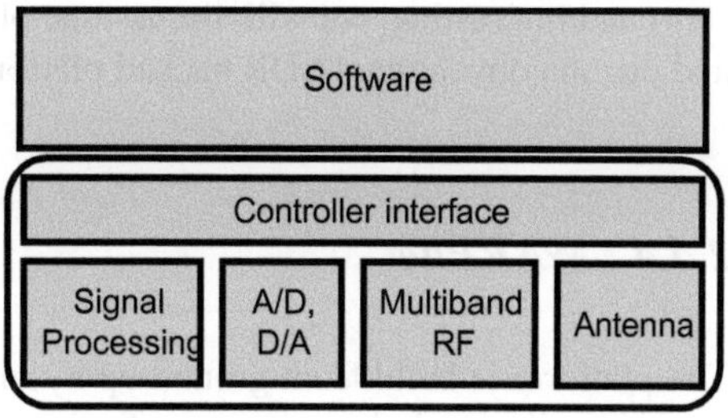

Fig. 2.4 Application software on top of SDR

2.4.3.1 GNU Radio

GNU Radio is a free, open source development toolkit for SDR [28]. It is licensed under GNU GPL version 3. It is widely used in academia and in commercial ecosystem. The toolkit comprises of different modules each representing an individual step in signal processing. The overall functionality of the toolkit is achieved by applications which renders themselves as a 'flowgraphs'. These flowgraphs interconnect different signal processing modules via a vertex which originates from a source module and terminates at a sink module, thus representing the flow of the data. The application flowgraph can be written in either C++ language or python language, while the core signal processing modules are written in C++ language. The GNU Radio comes with a graphical UI and called as GNU Radio Companion. The GNU Radio Companion provides drag-n-drop functionality to configure a particular application, thus allowing a novice user without having any programming knowledge to utilize this toolkit. GNU Radio is natively available for Linux, but other alternate build methods are available for Windows and MAC OS.

2.4.3.2 Pothosware

Pothosware is an open source software for interconnected processing blocks [29]. It provides multiple toolkits for different purposes. Pothos provide toolkit for interfacing with SDR hardware, provides another toolkit for real-time signal analysis and multiple toolkits signal processing blocks. Pothos also allows to integrate different components from other SDR softwares into the pothosware. The Pothos SDR toolkit provides a SDR source block for receivers and SDR sink block for transmitters. For interfacing with different hardwares, the Pothos project wraps SoapySDR project. SoapySDR is an open source C/C++ library for interfacing with different types of SDR hardware. The library is vendor neutral, implying that it is compatible with different types of hardware.

There exist multitude of software defined radio platforms, including Universal Software Radio Peripheral (USRP), eFalcon, Microsoft Research Software Radio (Sora), Cognitive Baseband Reconfigurable Radio (COBRA), Berkeley Emulation Engine (BEE2), and Wireless open-Access Research Platform (WARP). To keep it short and interesting, we will discuss one platform from traditional SDR: *WARPnet* and one unconventional SDR hacked platform: *RTL-SDR*.

2.4.4 WARPnet

The platform is build from ground up by integrating individual blocks at the lower layer to realize a whole cognitive radio functional platform [30]. The platform allows the developers to fine control the lower layer functionalities while allowing conventional users to deploy the platform for field testing. It utilizes two wide band radio boards to support desired transmission characteristics. One is MAX2829 which is dual band 2.4 and 5 GHz radio, while the other radio utilized is AD9352 which is a 2.5 GHz band radio. The radio board utilizes multiple dedicated control interfaces (channels) for coordination between different boards.

The WARPnet board utilizes a Xilinix Virtex-4 FX series FPGA. This FPGA provides the processing power required for doing complex computation over software for the working on real-time algorithms for wireless communication. The board has Marvell Alaska 88e1111 Gigabit Ethernet transceiver for bridging wireless to wired network infrastructure. The board is capable of supporting 2 GB of RAM. The board also has two 6.5 Gbs SATA bus for connecting hard drives for data storage. The board also support other hardware features which are not discussed in the book. Readers are suggested to go through the WARPnet documentation to get the detailed specifications.

Maxim MAX2829 RF transceiver provide analog baseband interfaces and handles RF translations. The transmitter has 16 bit dual input Analog Devices' AD9777 DAC. The receiver has dual 14bit output AD9248 ADC. The transceiver provide software control over transmit power, input and output filter etc. via SPI interface.

Analog device's AD9352 transceiver chip integrates the ADC and DAC along with the RF transceiver, filters and amplifiers. A dedicated Spartan-3 AN FPGA is utilized for providing control and data interface along with the processing algorithms for radio.

The board also utilizes a backdoor board which provides dedicated control channel and a control hub. This allows as a central controller to control all boards in the network. This includes remote programming of deployed boards in the network. The backdoor interface utilizes Ethernet and 900 MHz long range wireless interface for communication. The board utilizes GPS module to synchronize time with other boards for the purpose of MAC layer algorithms.

WARPnet architecture utilizes three functional cores: FPGA logic, the PowerPC and the Axis Etrax SoC. The high sped FPGA implements the physical layer of the cognitive radio via utilization of complex designs in HDL. The MAC layer is implemented in C/C++ language via the PowerPC as it has direct access to the lower physical layer. The overall cognitive control of the WARPnet node is done via the Axis Etrax SoC which provide control channel and monitoring of the network. In a WARPnet network, each node is connected and managed by a central entity called as WARPnet controller. The WARPnet controller monitors the system characteristics and provides a system wide view of the whole deployment. This allows easy calculation of the performance of the grid.

2.4.5 RTL-SDR

RTL-SDR is a software defined radio that is commonly found in the DVB-TV tuner dongle which utilizes Realtek RTL2832U chipset [31]. It was found that the RTL2832U chipset allows transferring raw I/Q samples to the host, thus allowing demodulation to be done in the software. The DVB-TV tuner is a cheap dongle which acts as a receiver, therefore with proper tools it can be utilized as a SDR receiver. The dongle utilizes a combination of a tuner, sampler and USB interface modules. The RTL2832U supports tuners at Intermediate Frequency of 36.125 MHz, low-Intermediate Frequency of 4.57 MHz, or Zero-Intermediate Frequency.

The RTL2832U provides 8 bit ADC sample with maximum theoretical sampling rate of 3.2 MS/s along with USB interface for pumping data to the host machine. Although the maximum bandwidth is 3.2 MHz, there are sample losses. The largest sample rate successfully without sample loss is 2.8 MHz. Different tuners provide different types of signal reception in different range of frequencies. Table 2.5 provides the list of tuners which is usually utilized with RTL2832U [32].

Although this hardware does not provide any transmission capability, it provides sensing capability which can be utilized as a starting point for people entering into cognitive radio field. The hardware does not provide accuracy to industry standards, but with proper noise/EMI protection, heat sinks and stable oscillator, a high level of accuracy can be obtained. The librtlsdr library provides generic driver interface for RTL2832U chipset [32]. Therefore most of the software defined radio software can be interfaced with this dongle via the help of librtlsdr library. The applications include FM radio, GSM listening, ADSB receiver, GPS receiver, AIS receiver etc.

Table 2.5 Different tuners with supported frequency bands

Tuner	Frequency band
Rafael Micro R820T	24–1766 MHz
Rafael Micro R828D	24–1766 MHz
Fitipower FC0013	22–1100 MHz
Fitipower FC0012	22–948.6 MHz
FCI FC2580	146–308 MHz and 438–924 MHz (gap in between)
Elonics E4000	52–2200 MHz with a gap from 1100 to 1250 MHz

2.5 Discussion

In this chapter various cognitive radio network technologies for QoS provisioning and energy management related works were reviewed and background materials were described which are essential to appreciate the following chapters. First, various spectrum management strategies were discussed and requirements of media access control techniques were introduced. Towards the end of the chapter, various energy management techniques were also discussed. Further a brief overview on different cognitive radio platforms was also described. The capabilities of these platform indicates that realization of the cognitive radio technologies. In the following chapters we focus exclusively on a framework which provides QoS provisioning and energy management for cognitive radio network, encompassing the technologies of spectrum management, media access scheme, energy management and also self-coexistence.

References

1. FCC. (2003). Revision of parts 2 and 15 of the commissions rules to permit unlicensed national information infrastructure (U-NII) devices in the 5 GHz band. *Memorandum opinion and order* (Vol. 21).
2. Atakan, B., & Akan, O. B. (2007). Biologically-inspired spectrum sharing in cognitive radio networks. In *2007 IEEE Wireless Communications and Networking Conference* (pp. 43–48).
3. Eletreby, R. M., Elsayed, H. M., & Khairy, M. M. (2014). Optimal spectrum assignment for cognitive radio sensor networks under coverage constraint. *IET Communications*, *8*(18), 3318–3325.
4. Niyato, D., Member, S., Hossain, E., & Member, S. (2008). Competitive pricing for spectrum sharing in cognitive radio networks: Dynamic game, inefficiency of nash equilibrium, and collusion. *IEEE JSAC*, *26*(1), 192–202.
5. Song, Y., Fang, Y., & Zhang, Y. (2007, November). Stochastic channel selection in cognitive radio networks. In *Global Telecommunications Conference, 2007. GLOBECOM '07. IEEE* (pp. 4878–4882).
6. Talat, S., & Wang, L.-C. (2009). Qos-guaranteed channel selection scheme for cognitive radio networks with variable channel bandwidths. In *International Conference on Communications, Circuits and Systems, 2009. ICCCAS 2009* (pp. 241–245).
7. Cai, L. X., Liu, Y., Shen, X., Mark, J. W., & Zhao, D. (2010). Distributed QoS-aware MAC for multimedia over cognitive radio networks. In *Global Telecommunications Conference (GLOBECOM 2010), 2010 IEEE* (pp. 1–5).
8. Wang, P., Niyato, D., & Jiang, H. (2010). Voice-service capacity analysis for cognitive radio networks. *IEEE Transactions on Vehicular Technology*, **59**, 1779–1790.
9. Timmers, M., Pollin, S., Dejonghe, A., Van der Perre, L., & Catthoor, F. (2010). A distributed multichannel MAC protocol for multihop cognitive radio networks. *IEEE Transactions on Vehicular Technology*, *59*, 446–459.
10. Cordeiro, C., & Challapali, K. (2007). C-MAC: A cognitive MAC protocol for multi-channel wireless networks. In *2nd IEEE International Symposium on New Frontiers in Dynamic Spectrum Access Networks, 2007. DySPAN 2007* (pp. 147–157).
11. Zhao, Y., Song, M., Xin, C. (2013). FMAC: A fair MAC protocol for coexisting cognitive radio networks. In *INFOCOM, 2013 Proceedings IEEE* (pp. 1474–1482).
12. IEEE 802.11 Standard (2012). IEEE 802.11 working group, wireless LAN medium access control (MAC) and physical layer (PHY) specifications.

13. Le, L., Hossain, E. (2008). A MAC protocol for opportunistic spectrum access in cognitive radio networks. In *Wireless Communications and Networking Conference, 2008. WCNC 2008. IEEE* (pp. 1426–1430).
14. Mishra, V., Tong, L. C., Chan, S., & Mathew, J. (2011). MAC protocol for two level QoS support in cognitive radio network. In *2011 International Symposium on Electronic System Design (ISED)* (pp. 296–301).
15. Wang, S., Wang, Y., Coon, J. P., & Doufexi, A. (2012). Energy-efficient spectrum sensing and access for cognitive radio networks. *IEEE Transactions on Vehicular Technology, 61*, 906–912.
16. Maleki, S., Pandharipande, A., & Leus, G. (2011). Energy-efficient distributed spectrum sensing for cognitive sensor networks. *IEEE Sensors Journal, 11*, 565–573.
17. Liu, Y., Xie, S., Zhang, Y., Yu, R., & Leung, V. C. M. (2012). Energy-efficient spectrum discovery for cognitive radio green networks. *Mobile Networks and Applications, 17*(1), 64–74.
18. Gao, S., Qian, L., & Vaman, D. R. (2008). Distributed energy efficient spectrum access in wireless cognitive radio sensor networks. In *2008 IEEE Wireless Communications and Networking Conference* (pp. 1442–1447).
19. Zhang, H., Zhang, Z., Chen, X., & Yin, R. (2011). Energy efficient joint source and channel sensing in cognitive radio sensor networks. In *2011 IEEE International Conference on Communications (ICC)* (pp. 1–6).
20. Zimmermann, A., Günes, M., Wenig, M., Ritzerfeld, J., & Meis, U. (2006). Architecture of the hybrid MCG-mesh testbed. In *Proceedings of the 1st International Workshop on Wireless Network Testbeds, Experimental Evaluation and Characterization, WiNTECH '06* (pp. 88–89). New York: ACM.
21. Tan, K., Wu, D., Chan, A., & Mohapatra, P. (2010). Comparing simulation tools and experimental testbeds for wireless mesh networks. In *2010 IEEE International Symposium on a World of Wireless Mobile and Multimedia Networks (WoWMoM)* (pp. 1–9).
22. Thompson, J. J., Hopkinson, K. M., & Silvius, M. D. (2015). A test methodology for evaluating cognitive radio systems. *IEEE Transactions on Wireless Communications*, **14**, 6311–6324.
23. Masonta, M. T., Mzyece, M., & Mekuria, F. (2012). A comparative study of cognitive radio platforms. In *Proceedings of the International Conference on Management of Emergent Digital EcoSystems, MEDES '12* (pp. 145–149). New York: ACM.
24. Gustafsson, O., Amiri, K., Andersson, D., Blad, A., Bonnet, C., Cavallaro, J. R et al. (2010) Architectures for cognitive radio testbeds and demonstrators 2014; an overview. In *2010 Proceedings of the Fifth International Conference on Cognitive Radio Oriented Wireless Networks and Communications* (pp. 1–6).
25. Lotze, J., Fahmy, S. A., Noguera, J., Doyle, L., & Esser, R. (2008). An FPGA-based cognitive radio framework. In *Signals and Systems Conference, 2008. (ISSC 2008). IET Irish* (pp. 138–143).
26. Mitola, J., III, & Maguire, G.Q., Jr. (1999). Cognitive radio: Making software radios more personal. *IEEE Personal Communications, 6*, 13–18.
27. Kuon, I., Tessier, R., & Rose, J. (2008). FPGA architecture: Survey and challenges. *Foundations and Trends in Electronic Design Automation, 2*, 135–253.
28. Blossom, E. (2004). Gnu radio: Tools for exploring the radio frequency spectrum. *Linux Journal, 2004*, 4.
29. SoapySDR Pothosware. (2016) Retrieved April 04, 2016, from http://www.pothosware.com.
30. Gupta, S., Hunter, C., Murphy, P., & Sabharwal, A. (2009). Warpnet: Clean slate research on deployed wireless networks. In *Proceedings of the Tenth ACM International Symposium on Mobile Ad Hoc Networking and Computing, MobiHoc '09* (pp. 331–332). New York: ACM.
31. Wickert, M. A., & Lovejoy, M. R. (2015) Hands-on software defined radio experiments with the low-cost RTL-SDR dongle. In *Signal Processing and Signal Processing Education Workshop (SP/SPE), 2015 IEEE* (pp. 65–70).
32. RTL-SDR. Retrieved November 01, 2016, from http://www.rtl-sdr.com/about-rtl-sdr.

Chapter 3
QoS Provisioning and Energy Management Framework for CRN

The cognitive radio comprises of the following three important functional components to realize its operation as wireless communication device:

1. Sensing: Sensing is at the center of the cognitive radio platform. This is responsible for sensing the environment and providing opportunistic information to the cognitive radio platform. Sensing information can be utilized locally or it can be passed to other devices responsible for utilizing the spectrum information. Examples would include passing the information to peer cognitive radio device for distributed spectrum usage learning. Another example would be passing the information to a central server which is responsible for making decisions.

2. Acclimation: A cognitive radio based on the sensed environment adapts its operating parameters to suit the desired operating behavior. The operating parameters primarily includes PHY and MAC layer of a cognitive radio device, but other OSI layers may need adjustment depending upon the network behavior.

3. Transmission/Reception: Once the transmission parameters are adapted, the cognitive radio node communicates with its peer node. To achieve such communication every node in the cognitive radio system need to adhere to a common standard of operation. This is required for any communication system to function properly.

Since cognitive radio operates in a sporadic heterogeneous environment, QoS provisioning in such system faces various challenges. A framework for cognitive radio network is discussed which includes functional modules responsible for QoS provisioning. To achieve this, an observation over the macroscopic and microscopic behavior of resource demand including *bandwidth* and *latency* is required. Different frameworks have been proposed and implemented by researchers for cognitive radio networks [1–4]; However not many framework focuses on the issue of QoS provi-

V. Mishra et al., *QoS and Energy Management in Cognitive Radio Network*,
Signals and Communication Technology, DOI 10.1007/978-3-319-45860-1_3

sioning in cognitive radio network from the perspective of Layer 2 of the protocol stack.[1] This book will integrate different technologies responsible for QoS and there relationship with each other.

3.1 QoS Parameters

As described in previous section, quality of service is defined as the ability of a system to provide a guarantee over a minimum grade of service offered to an inbound/outbound traffic. The two important parameters of QoS are *latency* and *throughput*. These two parameters have correlation with respect to resource access [5]. The system has varying degree of latency if the channel access is correlated with the bandwidth resource.

3.1.1 Latency Versus Throughput

Multiple factors affect latency including *propagation delay, serialization delay, routing/switching delay, queue and buffer management delay* [6]. For a user with bandwidth requirement more than the resource available, the latency is caused by an additional parameter called as resource access delay. A resource access delay of a packet refers to the time period for which the packet was deprived of resource since it was next in line for transmission in the queue [7]. This type of delay depends upon the access scheme used and hence the QoS provisioning framework includes resource access scheme as one of its important ingredient component. Throughput on the other hand refers to the rate of successful message delivery over a media [8]. This is affected by the bandwidth and quality of the selected wireless channel and also on the access scheme utilized by the user. Therefore channel selection is also included as one of the important ingredient composition of QoS provisioning framework.

To realize the quality of service, packet categorization is done to implement access categories. As discussed in Sect. 1.5, four access categories are supported by the discussed framework as shown in Table 3.1. Voice data is provided highest priority while background data is provided lowest priority. Therefore any packet generated by any application will fall within one of these categories. These access categories have different priorities and hence different handling techniques. Packets from higher access category are not dropped as often as packets from lower access category. Since a user may run multiple applications each generating traffic of different priority,

[1]Layers of the protocol stack refers to the architectural layers of the Open System Interconnection (OSI) model. This book will utilize the term protocol stack to represent functional layers of the OSI model.

Table 3.1 Priorities of different packets

Access category	Description	Marker
Voice priority	Highest priority, Low latency	0
Video priority	Second highest priority, Video can buffer if needed	1
Best effort priority	If no QoS mentioned, or burst mode traffic like web surfing	2
Background priority	Lowest priority, print jobs, no strict latency	3

packets are identified using packet marking scheme in packet headers [9]. Through the marking of packets in the header, the protocols realize which packet belongs to which category.

3.1.2 Self-coexistence and Its Role in QoS

Since wireless channels are unguided media (except in the case of beamforming, which is not considered in this work), operation of multiple wireless networks in the vicinity of each other suffers from interference [10]. Self coexistence refers to the method of establishing a non-interfering coexistence with other homogeneous networks in the vicinity. To achieve this coexistence, multiple methods have been proposed and most of them require sacrificing a part of resource so that interference is minimized. These sacrifices result in deterioration of QoS in the network. Therefore a method is required to cater graceful degradation of QoS while mitigating the interference for self-coexistence. This yields in the composition of self-coexistence management functionality in the QoS provisioning framework for cognitive radio network.

3.1.3 Energy Management and QoS

For green communication and to prolong the life of a battery operated device, it is fundamental to reduce the overall power consumption of the wireless network. This is achieved by reducing the transmission power and by operating on channels which require lower transmission power. Reducing the transmission power reduces the SNR/SINR of the signal and hence the QoS. Similarly, operating on channels with lower transmission power with lower bandwidth results in bandwidth loss and hence a degradation in QoS. Therefore an energy management system is incorporated as an ingredient of the composition of framework for the purpose of graceful degradation of QoS.

3.2 QoS Framework for Cognitive Radio Network

The high level view of the discussed QoS framework at the Layer 2 of the protocol stack is shown in Fig. 3.1. The figure shows different blocks/units along with their inter-network connections depicting the relationship between the components and the type of interaction.

The discussed framework works alongside the normal communication path allowing the re-utilization of conventional protocol stack. The application domain consists of multiple applications including the applications requiring QoS support. These applications could be multimedia application, VoIP, gaming etc. This domain conveys the QoS requirements of the applications to the other units in the framework. QoS control unit (QCU) is the block which receives the information from the application domain regarding the requirement of the QoS. It conveys this information to the Channel selection and management unit (CSMU) as well as to the MAC protocol unit (MPU). Environmental and RF channel block provide the information about the external environment including channel availability, and this information is passed to the CSMU and also to the MPU. The Energy Management Unit (EMU) is used to make important decisions with the help of other units to conserve the energy of the user. Channel Selection and Management Unit (CSMU) based on the knowledge

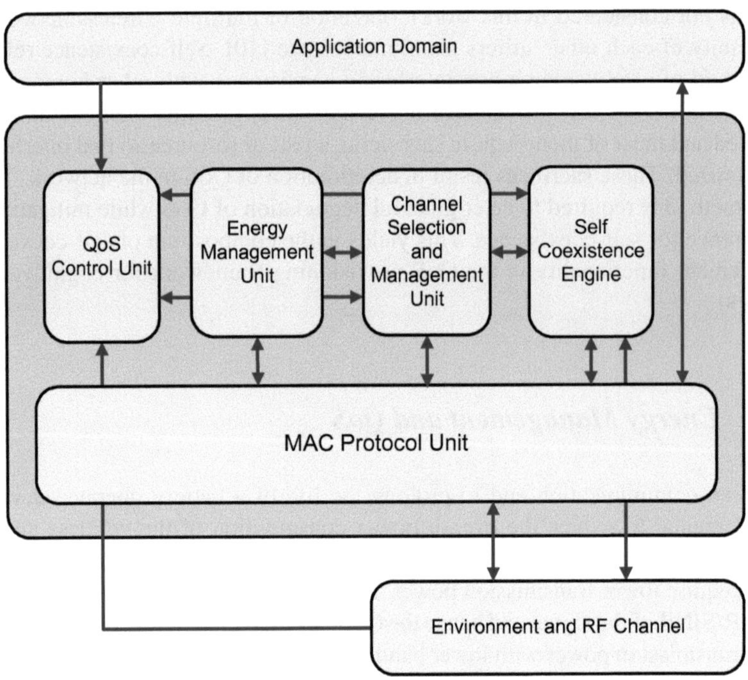

Fig. 3.1 QoS provisioning framework for cognitive radio network

from QCU select the channel which meets the required QoS level. Also, if required the CSMU utilizes the information from EMU to select a channel which will conserve the energy of the system. MAC Protocol Unit (MPU) considers all the information from all other units and selects/design appropriate MAC which conforms to all the requirements. If all requirements are not met, then an optimization step is required to achieve the desired level of result. A self-coexistence engine (SCE) is required to mitigate the issues of mutual interference experienced by multiple cognitive radio networks operating in the vicinity of each other. This unit interacts with CSMU and MPU to provide self-coexistence from the perspective of resource allocation and resource sharing respectively.

3.3 Detailed Layer 2 QoS Provisioning Framework

Figure. 3.2 shows the detailed view of the discussed framework for QoS provisioning at the Layer 2 of the protocol stack. The right portion shows the normal communication system path, in which the application sends the packets to the routing engine and from there it is sent to the MAC instance. A similar process happens in the reverse direction whenever a packet is received by the MAC instance. The left portion shows the QoS engine which works in parallel with the normal communication path. This portion receives the QoS requirement from the application (if application can provide) and performs actions needed to achieve the desired QoS with the help of different functional modules.

The framework is composed of different individual functional modules or components, which collectively operate towards achieving the common goal of network operation. The complete framework is interrupt driven and therefore any module would perform busy waiting until it receives an interrupt from source module. These modules are connected via function *call* and *response* method; parameters are passed via function calls and results are obtained in the form of response. Following are the description of different modules:

Application: Application is the user process generating traffic with varying priorities. User applications are assumed to generate 4 levels of priority traffic as shown in Table 3.1. Since a user can support multiple applications with varying levels of priorities, the application process allows multiplexing of different data types. An application can also act as a process for generating system traffic for the purpose of control plane.

Application QoS Engine: This component converts the application QoS requirement into the discussed framework language. For example if the application identifies the traffic communicated as VoIP traffic, application QoS engine will convert this information into the QoS bound value of the VoIP traffic. The QoS bounds are defined in terms of *latency* and *bandwidth*; and these values are stored in a database which are accessible to network operator. This component is integrated within the QoS Control Unit.

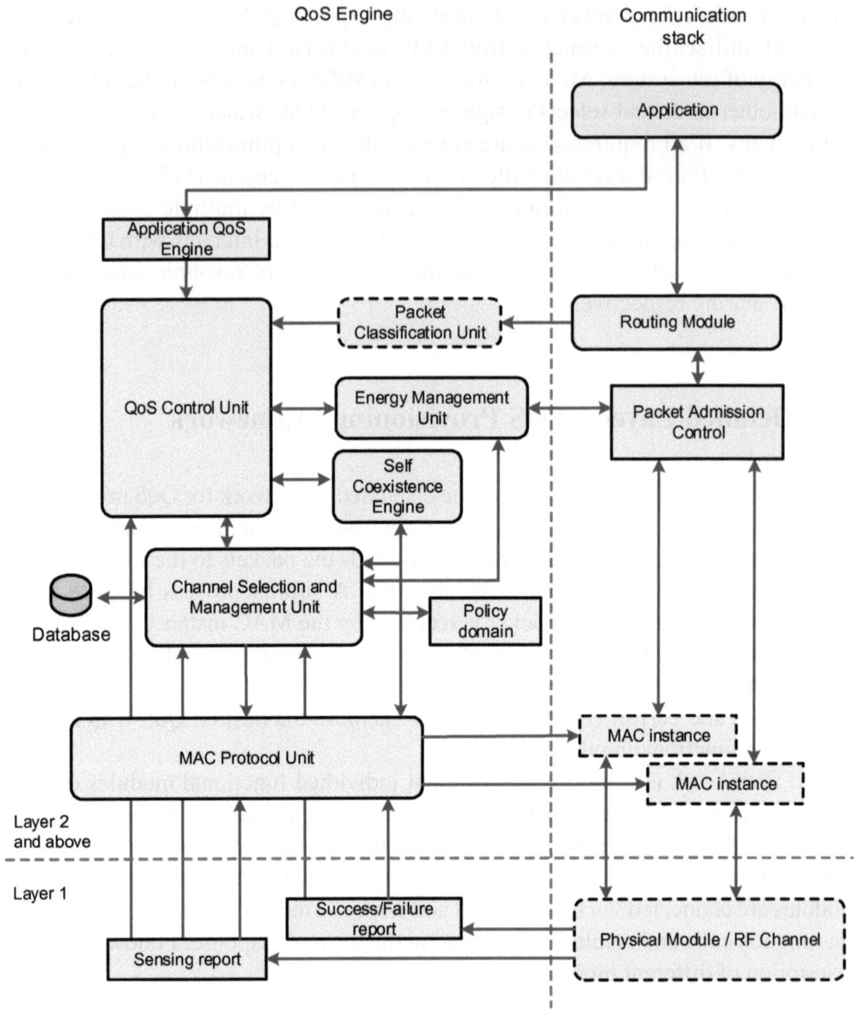

Fig. 3.2 Detailed QoS provisioning framework for cognitive radio network

QoS Control Unit: This unit is responsible for procuring and facilitating the QoS requirements of different traffic. Any other module/component may demand for QoS bound for the purpose of its operation. This unit is also responsible for managing and monitoring the experienced QoS of the system. QCU monitors the QoS of the system which it experiences through the utilization of current mode of the protocol stack with its current parameters. A feedback based control is utilized to manage the QoS by replacing/configuring the appropriate module which in turn will guide the QoS of the system between the defined bounds.

Policy Domain: Policy domain provides two sets of radio regulation policies: *local* and *global*, for a given geographical location. The local policy domain always supersedes the global policy domain. In case there is no entry in local policy domain for the given geographical location, global policy domain is accounted in. These radio regulation policies include allowed channels, maximum allowed power, operating bandwidth, channel separation, maximum usage time, sensing duration and sensitivity. Apart from providing these regulation policies, policy domain may also provide a list of optimal channels for selection. These optimal channels may be provided by third party spectrum monitoring institutes.

Channel Selection and Management Unit: Normally, channel selection is handled by PHY module (Layer 1) of a communication system, but in the discussed framework, this functionality is included at Layer 2 so as to enable close integration with other QoS related functionalities. The channels from Layer 1 are abstracted as identifiers to the Layer 2 along with their characteristics including bandwidth, occupancy statistics, SINR and transmission power requirement. Channel selection and management unit (CSMU) is responsible for selection of single or multiple wireless channels from a given list of channels. The channels are selected and facilitated to the media access scheme for communication purposes and therefore the number of channels selected depends upon the media access scheme. The given list from which the CSMU has to select channel is provided by policy domain. Various factors affect the selection of channel including occupancy, QoS constraint (latency and bandwidth) and energy consumption bound.

Energy Management Unit: Energy management unit provides the information about the remaining energy in the system. This unit helps in deciding whether a battery saving protocol should be activated or not. If a battery saving protocol is activated it provides information to the CSMU about the energy consumption bounds for the purpose of channel selection. In this way an optimized protocol can be used so as to minimize the energy consumption. Since the system tries to minimize the energy consumption, a graceful degradation in QoS is achieved with the help of packet admission control administered by energy management unit.

Packet Admission Control: Packet admission control unit is responsible for blocking/dropping certain number/category of data packets so as to provide resources to other higher priority data packets. This is mainly done when the total resource demand is greater than the available resource. Low priority packets are blocked in the queue and are transmitted when the resource available is commensurate to the demand. If resources are not available for prolonged time, the packets are dropped from the queue.

Traffic Classification Unit: Legacy applications may not have the functionality to explicitly provide the information about the priority of a traffic; this is overcome by providing a traffic classification unit. The work of traffic classification unit is to classify the incoming/outgoing traffic into one of the mentioned access category (Table 3.1). Different traffic classifiers utilize different methods for building up a classifier including statistical and rule based learning. Study of traffic classifier is beyond the scope of this work.

Routing Engine: Routing engine is a layer 3 technology which provides opportunistic routing protocols for packet routing. This unit is responsible for efficient routing, congestion control and energy efficient data transfer across the network. Routing engine may utilize information from other modules presented in this book. Since it is a layer 3 technology, it is beyond the scope of this book.

MAC Protocol Unit: This unit spawns/selects a MAC protocol which accommodates the requirements of QoS bounds of the given application. This unit gets the information about the QoS requirements from QCU. On receiving this information, the MPU retrieves the list of channels from CSMU and utilizes it to spawn a MAC instance. This unit is capable of generating multiple MAC instances for multi-radio system. In case of multi-radio system, all MAC instances operate in parallel by utilizing a multiplexing and demultiplexing scheme. This work does not consider the case of multiple radio.

Environment and RF Channel: This Layer 1 unit is equipped with radio which performs sensing and communication. Therefore reconfiguration of mixers, filters, modulator/demodulator and detector is done in this unit. Apart from reconfiguration of physical communication layer, this unit provides information about channel conditions like noise power, signal power, bit error rate etc. Since, this is a layer 1 technology, it is beyond the scope of this work.

Sensing Report: This unit encapsulates a spectrum detection functional module to provide the information about the availability of a channel. It receives information about channel parameters from Environment and RF module and facilitates CSMU for enhanced channel selection whereas the same information is utilized by an instance of MAC protocol to utilize/skip the frame. Sensing report unit is abstracted so as to present only the relevant information to the upper layers.

Success/Failure Report: This unit processes the information about the success/failure rate of the packets from the Environment and RF channel unit and pass it to the Channel Selection and management unit, so as to improve the decision on channel selections. The same information is passed to the MAC protocol unit which utilizes this information and if necessary, reconfigures its MAC instance on the fly. Again, this unit is abstracted so as to present only the relevant information to the upper layers.

3.3.1 Mode of Operation

Since cognitive radio is based on the technology of reconfiguration of protocols and communication parameters according to the environment in which it operates, the system operates with a particular set of parameters and protocols at a given time. This particular instance of protocol stack is termed as *mode* of protocol stack. In the discussed framework, a mode of protocol stack comprises of: *Network Mode*, *Spectrum Selection Mode*, *Media Access Mode* and *Self Coexistence Mode*. The protocol stack in such system is termed as multi-mode protocol stack. A network mode conveys the information about the mode of operation of the network: *infrastructure* or *ad-hoc*. Based on this mode the MPU unit reconfigures its selected *media* access mode for network operation. Spectrum selection mode provides the information about the spectrum selection technique (protocol) that is being utilized by the protocol stack. Similarly, self-coexistence mode provide information about the type self-coexistence protocol utilized by the protocol stack.

3.3.2 Generic Protocol Stack

The discussed framework utilizes different modules which share common functionalities and therefore a generic protocol stack can be introduced in control plane to build an efficient multi-mode capable system. To realize a generic protocol stack, the common functionalities are brought together to form a toolbox of protocol [11]. A generic protocol stack allows code/module sharing and accelerated protocol development using code re-usability. Therefore generic protocol stack has low memory footprint in comparison to other multi-mode protocol stack enabled by composition of different single mode protocol stacks.

Figure 3.3 shows the UML class diagram of a reconfigurable generic multi-mode protocol stack. As seen from the diagram, a generic multi-mode protocol stack com-

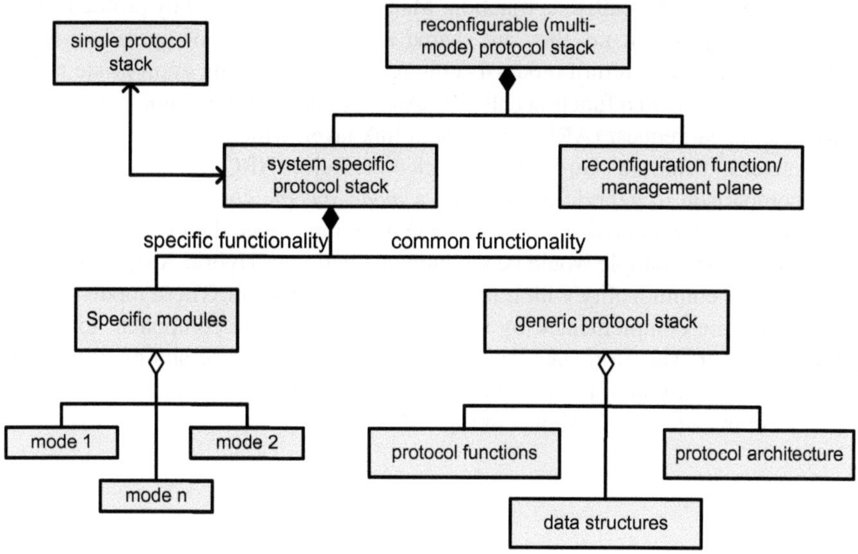

Fig. 3.3 UML diagram of the generic multi-mode protocol stack

prises of various elements that can be identified as analogous to each other in the context of communication protocols. Some of these commonalities include:

1. Functionality provided by a certain layer. An example would be parametrized function for selecting optional RTS/CTS mechanism in MAC layer.
2. Protocol architecture and framework, for instance structure of a MAC frame (sequence, duration etc.) and access rules may be shared across different MAC schemes.
3. Data structure or protocol data unit (PDU) utilized by a protocol can be utilized in a generic way across layer.

These common functionalities together with specific functionalities implements a system specific protocol stack. A reconfigurable multi-mode protocol stack is realized with the help of a reconfiguration function which parametrizes the generic and mode-specific modules. Reconfiguration function is realized via QoS control unit in management plane, which guarantees the convergence of fused complementary generic and mode-specific modules for spawning a system specific protocol stack.

3.3.2.1 Generic Versus Specific Partition

The partitioning of generic and specific functionality (system specific protocol stack or proprietary stack or specific modules) of a particular protocol is beyond the scope of this book. Nevertheless, a brief idea about the partitioning of functionalities is provided. In software engineering, a *generic* component refers to the identification of requirements that are met by a large family of abstraction but restrictive enough to efficiently realize its function. Many functions utilized by protocol stack are mode-independent, i.e. they are shared across all modes of operation. Some generic functions can return mode-specific results by providing appropriate mode-specific parameters in a function call [12]. An example would be implementation of automatic repeat request (ARQ) protocol in link layer, which can be configured by a parameter to behave as Go-back-N or Selective Reject ARQ.

In comparison to generic component, a *mode-specific* component refers to the function which is unique to a mode and cannot be implemented on shared basis with other modes. An example would be the implementation of Hybrid ARQ which does not share any commonality with traditional ARQ mechanism. These mode-specific components are complemented by their counterpart generic component to realize a protocol layer. Therefore each layer is composed of a generic and mode-specific part. A slice of particular instance of protocol stack with generic and mode-specific component in each layer is shown in Fig. 3.4.

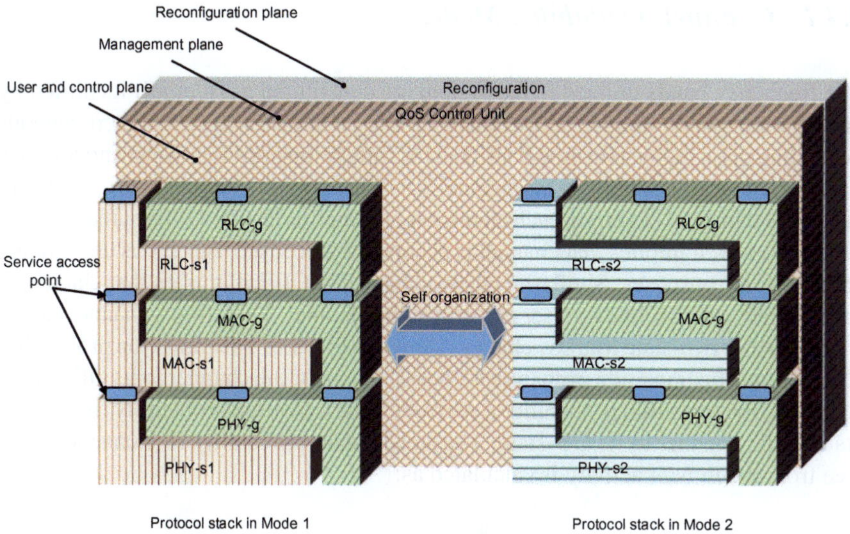

Fig. 3.4 Self organization of the protocol stacks between different modes of operation

3.4 Self Organization

The discussed framework can be configured to manually/automatically select a desired mode of operation of protocol stack. The *Network* mode is manually selected to operate a network in *infrastructure* or *ad-hoc* mode. All other *modes* can be configured to automatically select a desired mode of operation of the protocol stack so as to guarantee a minimum grade of quality of service. The self organization is done in management plane by QoS control unit. Therefore QCU continuously monitors the system overall performance in terms of latency, packet loss and achievable throughput.

Figure 3.4 shows a slice of protocol stack reconfiguring from mode 1 to mode 2 and vice-versa. The protocol reconfiguration happens at the reconfiguration plane which is controlled and managed by the management plane. Management plane functionality is realized by QoS control unit which continuously monitors the QoS of the system and reconfigures the protocol stack accordingly. The generic and mode-specific components are defined in user and control plane. Each layer of the protocol stack is connected to other layer via a service access point. Service access point acts as the interface for transfer of protocol data unit across layers. PHY-s1, MAC-s1 and RLC-s1 (Radio Link Control) represent the mode 1 specific components of physical, media access control and radio link control protocol respectively. Similarly PHY-g, MAC-g and RLC-g represent the generic component of physical, media access control and radio link control protocol respectively. These generic components are re-utilized in both modes as shown in Fig. 3.4. A self organization between different modes of operation can be initiated by management plane to achieve a certain grade of QoS in the system.

3.4.1 Channel Availability Model

The frequency bands utilized in the system are constituted as *channels* identified by an *ID* or a channel number. These channels can be homogeneous or heterogeneous depending upon the system model utilized for simulation purposes. Since modelling a licensed user traffic pattern on the channel is tedious in nature, the channel availability is modelled instead. To do so, each channel is assumed as an ON-OFF source. When a channel is ON, it implies that channel is occupied by licensed user and is unavailable for secondary purposes. Similarly, when a channel is OFF, it implies that channel is unoccupied and a secondary user can utilize the channel. Therefore a channel can be modelled as a two-state Markov chain as shown in Fig. 3.5. The transition probability α represents the transition probability from ON-state of the channel to its OFF-state. Similarly, the transition probability β represents the transition probability from OFF-state of the channel to its ON-state. The steady state probability of a channel to be free from incumbent activity is calculated as:

$$\Pi_i = \frac{\alpha_i}{\alpha_i + \beta_i} \; \forall i \in \{1, ..., C\} \tag{3.1}$$

Similarly, the steady state probability of a channel to be occupied by incumbent is calculated as:

$$\rho_i = \frac{\beta_i}{\alpha_i + \beta_i} \; \forall i \in \{1, ..., C\} \tag{3.2}$$

3.4.2 Quiet Periods

Standards designed by IEEE 802.22 WRAN group suggest that the maximum interference duration that a licensed user operating on a TV band can tolerate is 2 s [13]. Therefore according to the IEEE 802.22 standards, a cognitive access point should perform channel sensing every 2 s so as to ensure the protection of licensed incumbents. In the discussed framework, channels are sensed every 100ms. While sensing channels, it is made sure that all nodes (CR users) stop their communication for

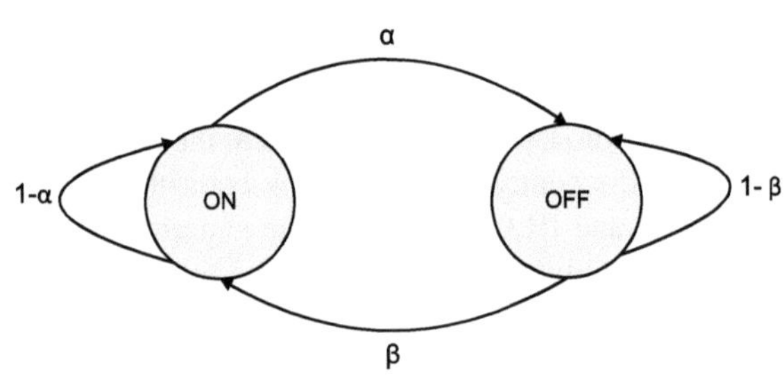

Fig. 3.5 Two state Markov chain availability model of a channel. ON state represent that channel is unavailable and vice-versa

the duration of sensing period. This duration is termed as *quiet period*. The design of quiet period follows the design of media access scheme. There are two types of sensing performed by a node: *in-band* and *out-of-band*.

3.4.2.1 In-Band Sensing

In-band sensing refers to the sensing performed by nodes on the channel in which they currently operate on. Since nodes normally exchange information with other nodes operating on the same channel, the distribution of quiet period for in-band sensing is easily obtained.

3.4.2.2 Out-of-band Sensing

Out-of-band sensing refers to the sensing performed by nodes in channels other than on which they are currently operating. Since sensing is performed in different channel, the distribution of quiet period on that channel needs to be obtained by the node performing out-of-band sensing. The design of information interchange about quiet period across all nodes in the network operating on different channels need to be considered while designing a media access scheme.

3.5 Discussion

Most of the research on QoS provisioning for CRN has been done individually without introducing a formal framework. Since cognitive radio is based on the technology of reconfiguration, a standard framework for CRN will allow easy and formal integration of different technologies responsible for QoS. The identification and compartmentalization of different technologies responsible for QoS will allow abstract and encapsulated behavior of framework design.

To achieve the above, a QoS provisioning framework for cognitive radio network at Layer 2 of the protocol stack is utilized. Different functional modules are identified for the purpose of QoS provisioning from the perspective of *latency* and *throughput*. While latency and throughput can be directly identified by the media access scheme and channel selection technique respectively, the effect of self-coexistence and energy management is separately studied. Macroscopic and microscopic functional modules at layer 2 of the protocol stack are discussed to provide an in-depth idea about the requirements. The concept of generic and mode-specific protocol stack is introduced to envisage the building blocks of the framework. From implementation and operation point of view, self-organization technique of multi-mode stack is also discussed.

References

1. Rondeau, T., & Bostian, C. (2009). *Artificial intelligence in wireless communications*. Artech House mobile communications series. Norwood: Artech House.
2. Ansari, J., Zhang, X., Achtzehn, A., Petrova, M., & Mahonen, P. (2011). A flexible mac development framework for cognitive radio systems. In *Wireless Communications and Networking Conference (WCNC), 2011 IEEE* (pp. 156–161).
3. Wysocki, T., & Jamalipour, A. (2009). Mac framework for intermittently connected cognitive radio networks. In *2009 IEEE 20th International Symposium on Personal, Indoor and Mobile Radio Communications* (pp. 481–485).
4. Rieser, C., Rondeau, T., Bostian, C., Cyre, W., & Gallagher, T. (2006). Cognitive radio engine based on genetic algorithms in a network. Jan 12 2006, US Patent App. 10/875,619.
5. Gamal, A., Mammen, J., Prabhakar, B., & Shah, D. (2004). Throughput-delay trade-off in wireless networks. In *INFOCOM 2004. Twenty-third AnnualJoint Conference of the IEEE Computer and Communications Societies* (Vol. 1, p. 475).
6. Doudou, M., Djenouri, D., & Badache, N. (2013). Survey on latency issues of asynchronous mac protocols in delay-sensitive wireless sensor networks. *Communications Surveys Tutorials, IEEE, 15*, 528–550.
7. Sakurai, T., & Vu, H. (2007). MAC access delay of IEEE 802.11 DCF. *IEEE Transactions on Wireless Communications, 6*, 1702–1710.
8. Bianchi, G. (1998). IEEE 802.11-saturation throughput analysis. *IEEE Communications Letters, 2*, 318–320.
9. Liu, C. (2007). Tracking DSCP marking of packets in a QOS enabled triple-play IP network. In *Third International Conference on Networking and Services, 2007. ICNS* (pp. 21–21).
10. Gardellin, V., Das, S., & Lenzini, L. (2013). Self-coexistence in cellular cognitive radio networks based on the IEEE 802.22 standard. *IEEE Wireless Communications, 20*, 52–59.
11. Sachs, J. (2003). A generic link layer for future generation wireless networking. In *IEEE International Conference on Communications, 2003. ICC '03.* (Vol. 2, pp. 834–838).
12. Siebert, M., & Walke, B. (2001). Design of generic and adaptive protocol software (DGAPS). In *Proceedings of 3G Wireless* (Vol. 1).
13. Cordeiro, C., Challapali, K., Birru, D., & Sai Shankar, N. (2005). IEEE 802.22: The first worldwide wireless standard based on cognitive radios. In *2005 First IEEE International Symposium on New Frontiers in Dynamic Spectrum Access Networks, 2005. DySPAN 2005* (pp. 328–337).

Chapter 4
Case Study: Spectrum Management in CRN Framework

Channel selection in the QoS provisioning framework (as discussed in Sect. 3.2), is done through channel selection and management unit (CSMU) along with the help of QoS control unit (QCU), policy domain and database as shown highlighted in Fig. 4.1. The CSMU module receives the requirement of QoS from QCU, and through the assistance of database and policy domain it selects channel. The selected channel identifiers are then passed to the MAC protocol unit (MPU).

4.1 Spectrum Usage Behavior

As discussed in Sect. 1.6, channel selection technique depends upon various spectrum characteristics including channel availability, SINR, bandwidth, path loss and primary user behavior. The primary user behavior in a cognitive operation based open spectrum system can be classified as: *deterministic* and *stochastic* usage behavior.

4.1.1 Deterministic Usage Behavior

There are some frequency channels which inherently follow a specific pattern of their usage over time because of the nature of the operating application [1]. This is verified by the spectrum utilization studies done by various researchers across the world, that have shown that the spectrum usage by incumbents follow a deterministic time pattern over space as discussed in Sect. 1.3 and work done in [2–18]. Spectrum occupancy measurement of Singapore is done in [19]. From this measurement and analysis, a table is drawn and shown in Table 4.1 for some channels whose observed spectral opportunity shows a deterministic time pattern.

V. Mishra et al., *QoS and Energy Management in Cognitive Radio Network*,
Signals and Communication Technology, DOI 10.1007/978-3-319-45860-1_4

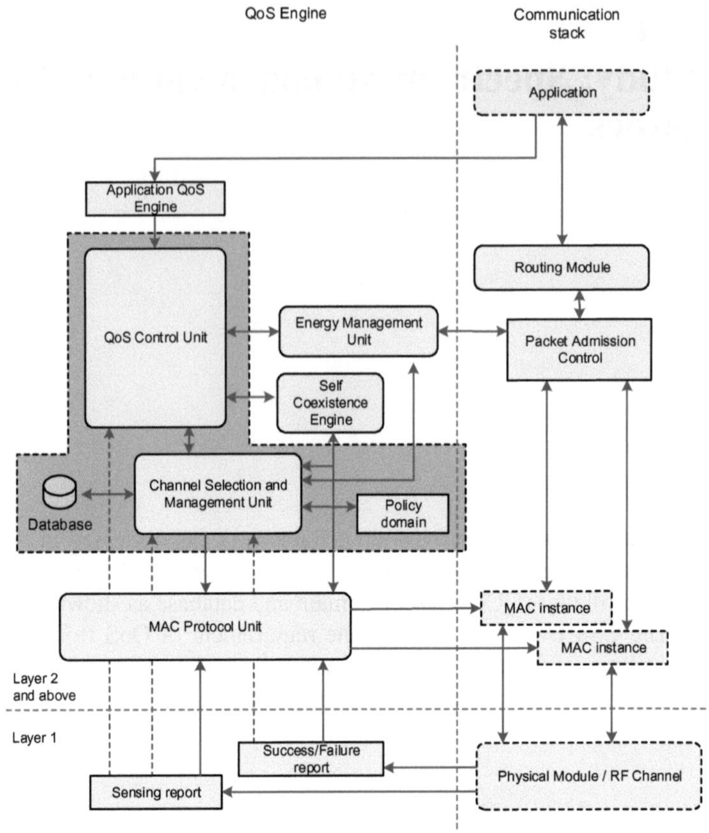

Fig. 4.1 Functional modules responsible for channel selection in the QoS provisioning framework for cognitive radio network

From Table 4.1 it can be observed that there are certain frequency bands which are always free during certain duration of time. Local TV channels like Suria (ch12) and Central (ch24) are closed from 12 am to 9 am and 12 am to 6 am respectively. Therefore these channels are well suited for cognitive radio communication during the mentioned time periods.

4.1.2 *Stochastic Usage Behavior*

If the spectrum usage pattern by a primary user does not follow a deterministic pattern, it is modelled with the help of statistical learning techniques to find the hidden pattern and to obtain the usage distribution. Several spectrum decision techniques have been proposed in various literatures (discussed previously) and are discussed in Sect. 1.6.

Table 4.1 Available frequency band over time in Singapore

Frequency (MHz)	Type	Free time (H)
80–88	PMR	18–12
183–186	Ch6	12–10
207–209	Ch9	18–16
212–214	Ch10	18–16
218–222	Ch11	11–06
225–230	Ch12	00–09
490–500	Ch24	00–06
614–620	Ch39	00–15
720–740	Channel 56–58	08–00
740–780	Channel 59–64	All Time
1880–1910	DECT and TDD	02–17
2270–2300	Fixed/Mobile	19–07

These works are used to model the spectrum characteristics with the assumption of stochastic usage behavior of incumbent usage pattern on the channel. Generally these schemes utilize the local and statistical observation to make a spectrum decision.

4.2 Reconfigurable Channel Selection

Recently, Federal Communication Corporation (FCC) has mandated the use of database along with spectrum sensing to observe the spectral opportunity at a given place over a given period of time [20]. The database assisted spectrum sensing provide an accurate measurement of spectrum characteristics. Since the spatial and temporal characteristics of spectral opportunities vary rapidly, an up-to-date incumbent operation database is needed for protected broadcast operation in a given location. Since providing an up-to-date information about primary service operation is exhausting in nature, another paradigm for information collection need to be determined. The channels can be classified into two types depending upon their usage: *Deterministic time pattern channels* and *Stochastic channels*. Therefore the system comprises of two step reconfigurable channel selection scheme. During the first step, a channel with deterministic time pattern is selected for usage. If no such channel exists, then stochastic channels are selected based on the requirements. The selection of stochastic channels can be assisted with the help of database.

Figure 4.2 shows the flowchart followed by the CSMU module of the discussed QoS provisioning framework along with the three main classes of channel selection: *Deterministic channel selection*, *Database assisted selection on the go* and *Selection on the go*. Initially channel selection starts with deterministic channel selection scheme, but if the observed spectral opportunities do not verify with that of database,

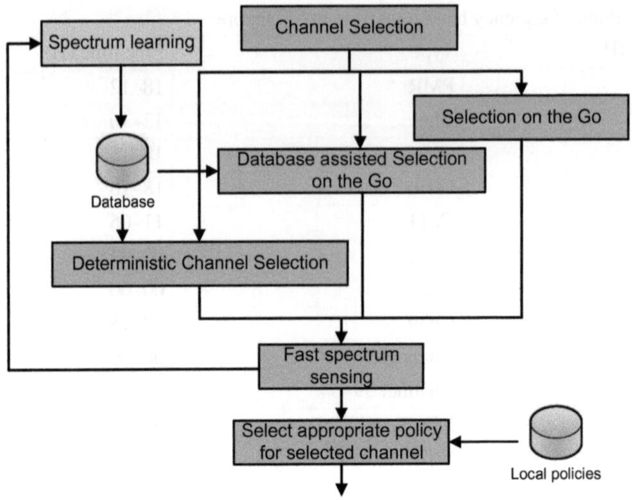

Fig. 4.2 Channel selection flow

then database assisted selection on the go method is used. In this method, statistically learning of spectral opportunities for stochastic channel is used along with database. If utilization of database in this methodology results in high error, then statistical learning is used without usage of databases which is called as independent selection on the go approach. Once a channel is selected, it is fast scanned to check out any incumbent activity, and if it is vacant then appropriate policies are selected for the channel. If the channel is not vacant, then deterministic spectrum learning module updates the database. An error parameter is utilized to realize the reconfiguration of channel selection schemes.

4.3 Channel Selection in Deterministic Environment

As discussed earlier, some channels have inherent characteristic of deterministic time pattern of utilization. These channels are allocated ahead of time for secondary open spectrum system so that better spectrum decisions can be made. A deterministic channel selection scheme for channel allocation ahead of time is discussed in [21]. Some of the spectrum decision criteria discussed for QoS provisioning are: *minimal channel switching* and *maximal throughput* of the system [22]. For the purpose of same, three different channel selection techniques to support minimal channel switching, maximal throughput and a value in between are discussed.

Channel selection techniques for deterministic environment employ two phases for its operation. The first phase involves learning of the channel spectral opportunity for 24 h and then based on the data obtained, the system may select a channel satisfying the given constraints. These constraints could be minimal channel switching for reducing the packet delay and losses, high throughput or a value in between. Users can also impose their own constraints to obtain the set of desired channels. The second phase involves usage of these channels and recording the errors in channel spectral opportunities. The errors are then used to check whether the system should continue with the current approach or change to the selection on the go approach.

4.3.1 System Model

The system comprises of N cognitive end-users coordinated with the help of a central controller *Cognitive Access Point* (CAP). Each user is equipped with a single half-duplex transceiver, capable of performing channel sensing. The system utilizes C non-overlapping heterogeneous primary bands each having bandwidth of B_c where $c \in C'$ and $C' = \{1, \ldots, C\}$. These C channels are selected from a set of TV and FM bands and are modeled as two state Markov chain as discussed in Sect. 3.4.1. A common control channel is utilized by CAP to coordinate with its users in the network; for which any ISM band or a channel with high a availability can be selected. An efficient spectrum sensing scheme from the literature is used at the physical layer for spectrum sensing; and it is assumed that the result obtained by it are correct. Therefore each CR user keeps track of the list of the available channels. This information is passed to the CAP on common control channel.

Each CR user node uses linked list as a database for capturing these spectral opportunities. There is a linked list assigned for each channel, therefore there are in total C linked lists maintained by each node. The nodes of each linked list contain the information of spectral opportunity represented by two fields: the *start time* and the *end time* (in time-units) during which the channel is free. A typical linked list is shown in Fig. 4.3.

The fields have the following meanings:
S_t: Start time from which the channel seems to be unoccupied by the primary user.
E_t: End time at which the channel ceases to be unoccupied by the primary user.
The subscript t represents the node ID within a linked list and is given by $t \in \{1, \ldots, T\}$ and T is the last node ID in the linked list. The value of T may differ for each linked list.

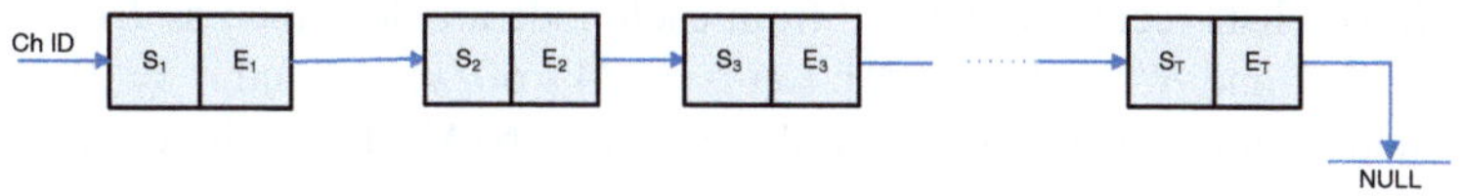

Fig. 4.3 Linked list

4.3.2 Deterministic Learning with Spectrum Selection and Usage

The system samples all C channels for 24 h with sampling interval of x minutes and sampling time of z seconds (where $z << x * 60$) i.e. the channel will be sensed every x minutes for z seconds. Thus the total time required for sampling C channel is $z * C/60$ min (where $z * C/60 < x$). The value of x is chosen such that the system would not sample channels too frequently nor miss the important transitions on the channels. The values of S_t and E_t in Fig. 4.3 represent the time-units mapped to the real world time with $(S_t, E_t) \in \{0, \ldots, 1439\}$/sampling interval; whereas 1 time-unit duration = sampling interval (x minutes). The terms time-unit and time-slot are used interchangeably. If x is small, then the number of nodes in the linked list may increase, and in the worst case scenario, the maximum number of nodes in the linked list equals $((24 * 60)/x)/2$ when the channel alternates between free and occupied states in consecutive samplings. To shorten the length of a linked list, a threshold parameter Ω is utilized. If the length of the spectral opportunity is greater than Ω (i.e. $\forall t\{E_t - S_t \geq \Omega\}$), then this spectral opportunity is inserted in the linked list. For example if $\Omega = 20$, then if one channel is sensed free for 19 consecutive time-units while other channel is sensed free for 22 consecutive time-slots, the availability information of the former channel will not be entered in the linked list while that of the latter channel will be entered.

At startup, all linked lists are empty and point to address NULL. With each sampling interval and satisfaction of the threshold parameters, the nodes are updated or inserted in the linked list. The time-unit starts from 0 (00:00 h) and increases by 1 with each sampling interval or every x minutes. An example of the linked list of Channel 2 after 24 h using x=10 min is shown in Fig. 4.4. It can be observed that the Channel 2 is unoccupied during the time periods 00:00–11:40 h, 16:40–20:10 h and 23:00–24:00 (00:00); and is occupied during the time periods 11:40–16:40 h and 20:10–23:00 h.

The sampling continues for 24 h, at the end of which the system would have obtained spectral opportunities for all channels. This information is passed to CAP, which consolidates the information received from all the nodes to obtain the global system spectral opportunity. This consolidation is required so that every node in the system agrees on the availability of the channel. A simple method to obtain the global channel opportunity is to convert all the linked lists into matrices and compute the intersection on them, thus reducing the computation complexity.

The linked lists received by CAP from each node are described as $LL_{i,j}$ where $i \in C'$, $j \in N'$ and $N' = \{1, \ldots, N\}$. Let $\eta = (24 * 60/x) - 1$ and $\eta' = \{0, \ldots, \eta\}$. Let the matrix for the linked list received from node j be $\mathbf{MAT}(j)$, where $j \in N'$. The

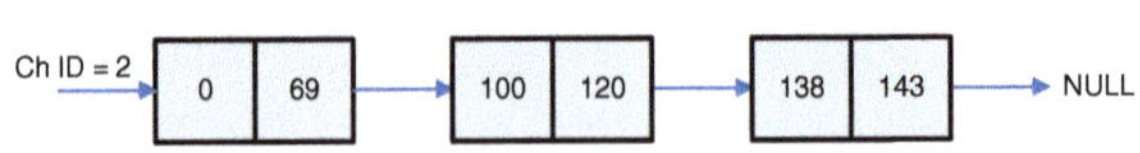

Fig. 4.4 Channel usage pattern of Channel 2 for one day (24 h)

row and column of the matrix represent channels and time-units respectively. The conversion from linked lists to this matrix can be done as follows. In the following conversion index j is dropped.

$$\mathbf{MAT_{i,k}} = \begin{cases} 1, & \exists t[k \in \{S_t \dots E_t\}] \\ 0, & \text{otherwise} \end{cases} \forall i \in C', \forall k \in \eta', t \in T' \tag{4.1}$$

The predicate logic $\exists t[k \in \{S_t, \dots, E_t\}]$ or $\exists t[S_t \leq k \leq E_t]$, evaluates to true whenever the time-unit value falls within the node field S_t and E_t values. Therefore the elements of the matrix are either 0 or 1 representing a PU is present or not respectively on a channel corresponding to the given time-slot for the node j. Overall, N such matrices are obtained. To consolidate all the global information, the CAP performs AND operation on all the corresponding locations of the matrices to obtain a channel availability matrix satisfying the channel availability of all the nodes in the system, i.e.

$$\mathbf{M_{i,k}} = \bigcap_{j=1}^{j=N} \mathbf{MAT_{i,k}}(j), \forall i \in C' \text{ and } \forall k \in \eta' \tag{4.2}$$

The channel availability matrix **M** agrees with all the CR nodes present in the system with respect to the availability of the channel. An example matrix is shown in Fig. 4.5. It is assumed that at any given time-slot (time-unit) at least one channel is available for communication. Hence

$$\forall k \exists i \{\mathbf{M_{i,k}} = 1\}, \text{ where } k \in \eta', i \in C' \tag{4.3}$$

Different users have different requirements for the channel selection. The matrix **M** can be processed to obtain the desired channel or set of channels satisfying the users constraints. If a single channel is required at a given time-slot, then the solution is represented with a vector. The vector index is represented by the time-slot while its content represents the channel ID. Therefore this vector guides the system to choose a given channel at a given time-slot. A typical vector could be initialized as:

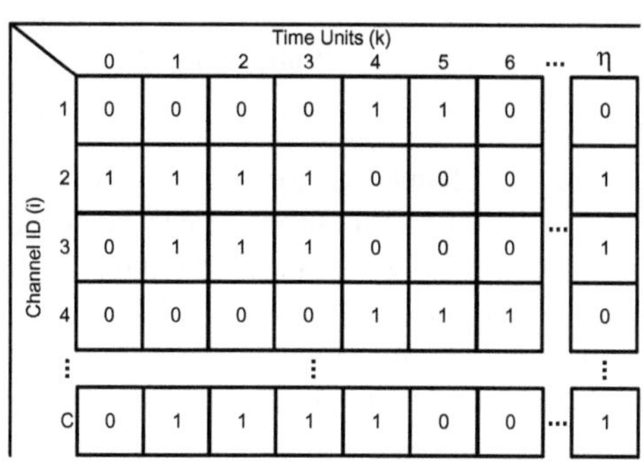

Fig. 4.5 A typical example of matrix showing availability of channels in CR network

$$\mathbf{V_k} = \{i : \forall k \exists i [\mathbf{M_{i,k}} = 1]\}, \text{ where } \forall i \in C' \text{ and } \forall k \in \eta' \tag{4.4}$$

Multiple vectors may satisfy this equation and hence multiple solutions may exist if no constraints are applied. The above vector simply chooses a channel which is available during a given time-slot and thus represents the solution space for all possible vectors. A user may apply single or multiple constraints on the system to impose restriction on channel selection. That is, the solution vector would be an element from the solution space set obtained in Eq. 4.4. Three different strategies of channel selection are discussed so as to meet the constraints of: *minimal channel switching, maximal throughput* and a *value in between.*

4.3.3 Minimal Channel Switch Requirement

Channel switching is costly for a radio device especially in CR devices [23]. Real-time applications suffer from packet losses, jitters and delays induced by frequent channel switching in CRN. Therefore a comparison of different channel switching techniques indirectly corresponds to a comparison of different packet delays, losses and jitters induced by channel switching in the network. Minimal channel switching refers to the selection of channels such that the system would stay on that channel for a long period of time without hopping onto other channels. The problem can be formulated as to obtain a solution vector **VMIN** such that when the system follows the solution vector, the overall switching is minimized for the entire duration. Therefore the problem is to find **VMIN** which has the least total number of channel switching as compared to other vectors available, i.e.,

$$\sum_{k=0}^{\eta-1} Switching(\mathbf{VMIN_k}) \leq \sum_{k=0}^{\eta-1} Switching(\mathbf{V_k}) \tag{4.5}$$

where $\forall V \in$ {set of all possible vectors $\mathbf{V}$ obtained from Eq. 4.4} and $Switching()$ is a binary function defined as:

$$Switching(X_i) = \begin{cases} 1, & \text{if } X_i \neq X_{i+1} \\ 0, & \text{otherwise} \end{cases} \tag{4.6}$$

where X is channel vector and i is its time index.

A solution to this problem can be obtained by using a greedy approach. This approach selects the channel which has the maximum period of continuous vacancy over a time interval. Algorithm 4.1 gives the implementation detail of the greedy approach. The algorithm first initializes the vector to value $zero$ (line 1). If the solution vector contains $zero$ at any time-slot, it means there are no channels available during that time-slot. The outer loop is used to index the time-slots (lines 3–18), the middle loop is used to check out all the channels (lines 4–14) and the inner most loop is used to check out the length of the continuous vacancy (lines 7–11). The variable array *trail* stores the length of continuous vacancy over all channels. At the end, another loop is used to check which channel has the longest spectral opportunity and is stored in **VMIN** (lines 15–17).

Lemma 4.1 *If on average there are θ channels available per time-unit, then the total number of vectors* **V** *available is $\theta^{\eta+1}$. The average is obtained over space or location.*

Algorithm 4.1: Algorithm for minimal channel switching using greedy approach. The *max*() function returns the maximum value among all elements of a vector and *argmax*() function returns the argument for which the function value is maximum.

```
1  VMIN_k ⟵ {0}, ∀k ∈ η′;
2  trail_i ⟵ {0}, ∀i ∈ C′;
3  for k = 0 to η − 1 do
4      for i = 1 to C do
5          trail_i ← 0;
6          if M_i,k = 1 then
7              for d = k + 1 to η do
8                  if M_i,d = 0 then
9                      break-for;
10                 end
11             end
12             trail_i ← d − k
13         end
14     end
15     for k = k to k + max(trail) do
16         VMIN_k = argmax_{j∈C′}(trail);
17     end
18 end
```

Proof The total number of vectors **V** equals to the total number of combinations through which the system can find at least one channel available in each time-slot. Since on average there are θ channels available at each time-slot, the total number of combinations available from time-unit $= 0$ to time-unit $= \eta$, equals $\theta.\theta.\theta \ldots \eta + 1$ times i.e. $\theta^{\eta+1}$. Thus, the search space is exponential.

Lemma 4.2 *The algorithm based on the greedy approach of selecting the longest consecutive free time-slots provides an optimal solution to the minimal channel switching problem. In other words there exist no other vector* **U** *that has smaller number of channel switching than that obtained by the Algorithm 4.1. The proof is given in Appendix A.*

4.3.4 Maximum Throughput Requirement

Some bandwidth intensive applications like peer-to-peer (P2P) file sharing require high throughput across 24 h. Normally, these applications are unconstrained from the perspective of channel selection and the only requirement is high bandwidth. Therefore, an un-constrained channel selection technique with the aim of maximizing throughput across 24 h is provided. Maximum throughput tends to provide channel selection strategy in such a way that the system will select those channels which provide highest bandwidth. Therefore the solution vector to this problem should provide the list of channels which would provide maximum bandwidth over a given time-slot. The maximum throughput vector **VTH** can be calculated as shown in Algorithm 4.2.

The algorithm first sorts the IDs of the channel in such a way that the lowest index has the highest bandwidth. The mapping is done internally from the older index to the newly assigned index, and is not shown here. The algorithm then selects the channel which has maximum bandwidth over a given time-slot and enter its ID into the vector **VTH**.

Algorithm 4.2: Algorithm for obtaining maximum throughput vector

Rearrange the rows representing the channels in the matrix M such that lowest index has the highest bandwidth i.e. $B_i > B_{i+1}, \forall i \in C'$;
VTH$_k \longleftarrow \{0\}, \forall k \in \eta'$;
for $k = 0$ **to** η **do**
    for $i = 1$ **to** C **do**
        if M$_{i,k} = 1$ **then**
            VTH$_k \leftarrow i$;
            break-for;
        end
    end
end

4.3.5 Intermediate Solution to Provide High Throughput Along with Minimal Channel Switch Requirement

The previous two approaches provide either the minimum number of channel switching or the highest throughput. Sometimes, however it is required to have a solution in between i.e. to provide a vector consisting of channel IDs such that if the system follows the list, then it would provide an intermediate solution which lies in-between the two approaches. Throughput loss caused by channel switching is not considered, as switching to a channel having a high bandwidth can easily compensate for the loss. Therefore the aim of this channel selection mechanism is to find a solution vector such that it provides bounded channel switching (for bounded packet loss, jitter and delay induced by channel switching) along with the guaranteed throughput. This problem lies in the domain of NP-complete problems, and called as the High Throughput Minimal Switching (HTMS) problem.

Complexity Analysis of HTMS Problem

The optimization problem of HTMS is converted into decision problem to define its complexity. Every optimization problem can be converted to a decision problem [24], and the decision problem corresponding to HTMS problem is defined as follows:

Definition 4.1 *Given a channel availability matrix* **M***, and numbers* ω *and* δ*, does there exist a vector* **V** *which satisfies all the constraints of the HTMS problem along with the condition that total number of switching* $<\delta$ *and bandwidth* $>\omega$*?*

Lemma 4.3 *HTMS decision problem is in NP. The proof is given in Appendix B.*

Lemma 4.4 *HTMS decision problem is NP-Hard. The proof is given in Appendix C.*

Theorem 4.1 *HTMS decision problem is NP-complete problem.*

Proof From Lemmas 4.3 and 4.4, it can be concluded that HTMS problem is NP-complete problem.

Heuristic Algorithm for HTMS problem

Since HTMS problem is NP-complete problem, a simple heuristic approach for its sub-optimal solution is provided in Algorithm 4.3. The algorithm first selects the minimal channel switching vector and then tries to satisfy the goal of guaranteed bandwidth. This is done by selecting the channels which were previously not selected and have higher bandwidth than that of the currently selected channel. The allowed switching and guaranteed bandwidth is represented by δ and ω respectively.

The algorithm first obtains the minimal channel switching vector **VMIN** from Algorithm 4.1. It then updates the vector with the channel IDs which have higher bandwidth than the currently selected channels in **VMIN** (lines 5–12). The *while* loop iterates until all channels have been checked (lines 14–16) or the bandwidth desired is obtained (line 3) and hence guarantees the convergence.

4.3.6 Complexity Analysis of Algorithms

The worst case scenarios for all algorithms are observed. Algorithm 4.1 has one parent *for* loop (lines 3–18) with two child *for* loops (lines 4–14 and lines 15–17). The running length of the parent loop is η and that of child loops are C and α, whereas the parameter α is variable and its value depends on the parameter η. The child loop (lines 4–14) with running length C is nested with another child loop (lines 7–11) with

Algorithm 4.3: Algorithm for intermediate channel selection for HTMS problem. The function *band_sum*() returns the overall bandwidth of the system.

```
1  Obtain VMIN from Algorithm 4.1
2  i = 1;
3  while ω ≤ band_sum(VMIN) do
4      flag = 0;
5      for k = 0 to η do
6          if M_{i,k} = 1 & VMIN_k > i then
7              VMIN_k = i ;
8          end
9          if ω ≤ band_sum(VMIN) then
10             flag=1; break-for;
11         end
12     end
13     i = i + 1;
14     if i = C || flag = 1 then
15         break-while;
16     end
17 end
```

running length α. The outermost parent loop (line 3) is complemented with the help of innermost for loop (line 7) and the other child for loop (line 15). Thus the running length of outermost loop is complemented. The complexity of $max()$ function and $argmax()$ function for current application is assumed to be $O(n)$, where n is the length of the vectors for these functions. Thus the complexity obtained of Algorithm 4.1 is $O(\eta(C+n))$ or $O(\eta^2)$.

Algorithm 4.2 has two nested for loops with running length of η and C. Thus the complexity of the algorithm is $O(\eta C)$ or $O(\eta^2)$.

Algorithm 4.3 obtains the vector **VMIN** using Algorithm 4.1 and then updates the vector to satisfy the bandwidth guarantee. The outer $while$ loop has running length of C and the inner for loop has running length of $\eta+1$. Thus the complexity of this algorithm equals, the complexity of Algorithm 4.1 $+O(C(\eta+1))$ or $O(\eta^2)$.

4.3.7 *Spectrum Usage*

When the system has obtained the spectral opportunities for 24 h and channel selection vectors (following the user constraints), the selected channel can be utilized for communication purposes. Spectrum usage consists of normal communication between nodes in CRN using the channels provided by the vectors obtained with different channel selection algorithms. An error counter ξ_i, $\forall i \in C'$ is associated with each channel. Each time before a packet is transmitted, fast sensing [25] is performed on the channel to detect the presence of any PU. If a PU is present on the channel, then that information is updated in the global channel availability matrix **M** and the error counter associated with that channel is increased by 1. If the error counter reaches the threshold Ψ, then that channel (row) information is removed from the channel availability matrix **M**. If the total error on all channels crosses the threshold χ, then the system switches onto Selection on the go approach. The parameters Ψ and χ are user selectable and depends upon the channel environment.

4.3.8 Performance Analysis

Since deterministic time pattern of the channels are being utilized, this model is initialized for the first 24 h of the system, after that the recorded state is used for the next 24 h. The given scheme is evaluated using discrete event simulation in Matlab which simulates the network up to frame level. The channel availability parameters (discussed in Sect. 3.4.1) α and β are set to 0.1, so as to model the channels whose availability do not change abruptly. A low value of α and β means that the system is likely to stay either in ON state or in OFF state most of the time without state transition. The bandwidth (data rate) of the channel is modelled as to be linear with the number of channels used, i.e. the minimum bandwidth of a channel is given a value of 2 Megabit per minute (Mbpm) and the maximum bandwidth of the channel is $2 * C$, where C is the maximum number of channels used for the simulation. Therefore the bandwidth is distributed linearly from Channel 1 to Channel C as 2 to $2 * C$ Mbpm. The error counter ξ_i, $\forall i \in C'$ is set to 10.

The discussed techniques are compared against three other channel selection techniques: stochastic channel selection (SCS) [26], automatic channel selection (ACS) [27], and minimum variance-based spectrum decision single selection (MVSD-SS) [23]. SCS [26] utilizes learning automata and aims at reducing the number of channel switching by selecting a channel which maximizes the probability of successful transmission. ACS [27] aims at selecting a channel that provides guaranteed bandwidth and fairness to the system with the help of measured primary user activity on the channel. MVSD-SS [23] aims at selecting a channel that increases the bandwidth utilization of the system. For a fair comparison, some of the constraints have been removed which were utilized in the literature. The given channel selection techniques utilize the infrastructure version of the media access scheme discussed in Chap. 5. Each data point in the graph is an average of 100 runs while the unit of bandwidth and sampling interval are Mbpm and minute respectively.

The algorithms is evaluated on the basis of following performances:

1. Overall Bandwidth of the system
2. Overall Channel switching required.

Overall bandwidth of the system refers to the bandwidth obtained by the system in 24 h. At some point the bandwidth obtained might be less and sometime might be the high. The bandwidths are consolidated by summing them up for 24 h. Overall channel switching required refers to the total switching needed for the 24 h operation of the system.

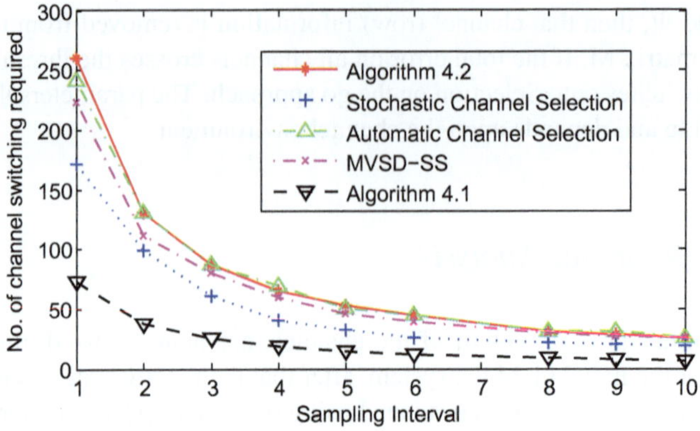

Fig. 4.6 Number of channel switching required versus sampling interval (x minutes)

Figure 4.6 shows sampling interval versus number of channel switching required for overall duration of 24 h with $C = 10$. It can be seen from the graph that Algorithm 4.1 provides minimal channel switching which is better than SCS technique. Although SCS technique aims at minimizing the channel switching, it lacks the knowledge of spectral opportunity ahead of time and thus suffers from degraded performance. All other techniques do not consider channel switching in their decision making process and therefore suffers a large amount of channel switching. Figure 4.6 also implies that Algorithm 4.2 is not good for minimal channel switching as it has the worst performance.

Figure 4.7 shows the bandwidth obtained versus sampling interval with $C = 10$. Results show that the bandwidth of the system is independent of the sampling interval used. It is also noted that Algorithm 4.2 provides the highest bandwidth among all

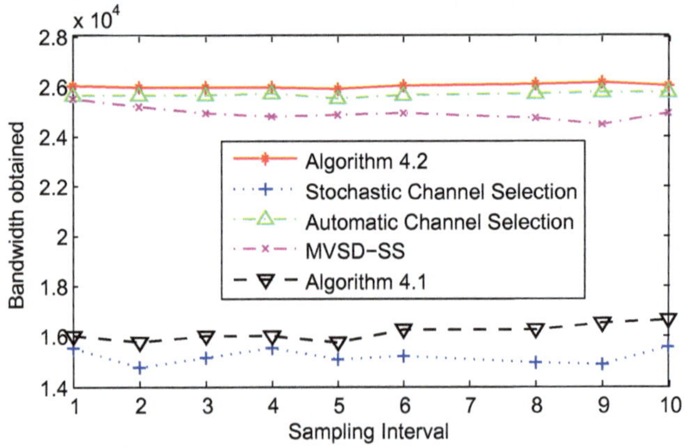

Fig. 4.7 Bandwidth versus sampling interval (x minutes)

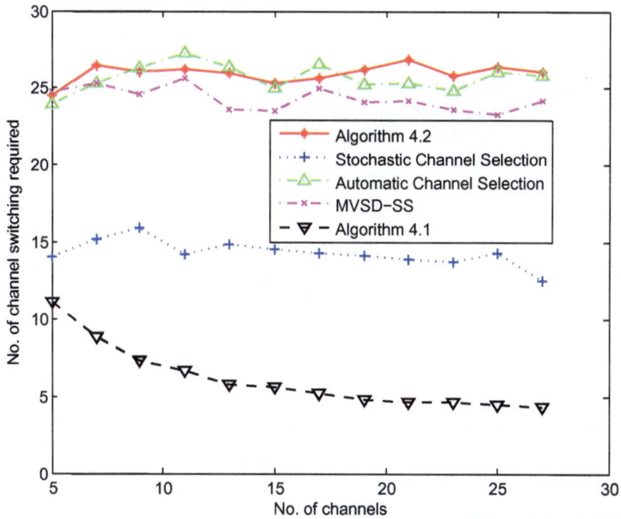

Fig. 4.8 Number of channel switching required versus number of channels utilized in the system

other techniques. This is the maximum unconstrained bandwidth that can be obtained by the system. Despite the removal of constraints from ACS and MVSD-SS to select high bandwidth channels for a fair comparison, these techniques do not have the knowledge of spectral opportunities ahead of time and suffer from performance loss. SCS aims at reducing the number of channel switching and does not account for throughput. Therefore its performance is near to that of Algorithm 4.1.

Figure 4.8 shows the number of channel switching required versus number of channels in the system with sampling interval $= 10$ min. It might be noted that the number of channel switching required in Algorithm 4.2 and all other techniques except Algorithm 4.1 is insensitive to the number of channels used in the system. The number of channel switching decreases with an increase in the number of channels in the system for Algorithm 4.1. With an increase in number of channels in the system, the number of combinations of vector $\mathbf{V}$ increases in the channel availability matrix and hence the switching decreases.

Figure 4.9 shows bandwidth obtained versus number of channels used in the system with sampling interval $= 10$. Note that Algorithm 4.2 obtains the highest bandwidth while other techniques (ACS and MVSD-SS) perform marginally poorer. The difference becomes significant when the number of channels used in the system becomes large. The performance of the Algorithm 4.1 and SCS technique is similar. The bandwidth of each scheme increases linearly because the bandwidth of each channel also increases linearly with the number of channels in the system (bandwidth is distributed linearly from channel 1 to channel C as 2 to $2 * C$ Mbpm).

Figure 4.10a, b show the performance for Algorithm 4.3 with guaranteed throughput (bandwidth) and maximum allowable switching. For the purpose of generalization, the rule of thumb "more than half" is used and assumed that the number of

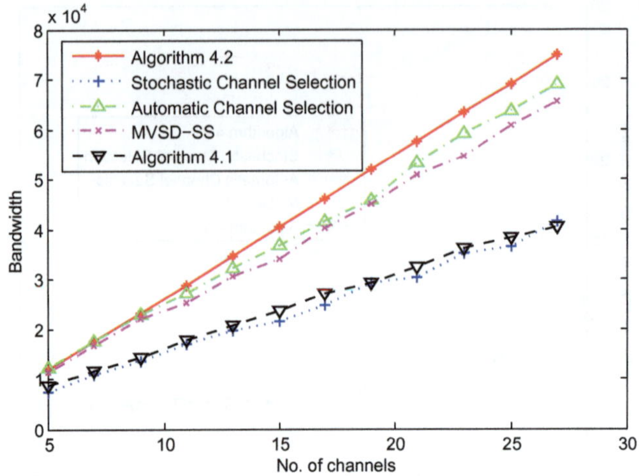

Fig. 4.9 Bandwidth versus number of channels

switching allowed per slot = 0.1 and total bandwidth available per slot = 60 % of the maximum bandwidth of the channel. The parameters α and β are set to 0.1.

Figure 4.10a shows number of channel switching allowed versus number of channels in the system with sampling interval = 10. The figure shows that the algorithm obtains channel switching lower than the allowed value. In addition the graph shows that there is no particular relation between the number of channels used in the system with the number of channel switching required.

Figure 4.10b shows bandwidth obtained in the system versus number of channels used in the system with sampling interval = 10. Note that Algorithm 4.3 satisfies the bandwidth guarantee required by the system. The bandwidth of each scheme increases linearly because the bandwidth of each channel also increases linearly with the number of channels in the system.

Figure 4.10c shows number of channel switching allowed versus sampling interval for an overall duration of 24 h with $C = 10$. Algorithm 4.3 provides channel switching lower than that of the allowed value. The allowed switching (δ) decreases with the increase in sampling interval value and so is the value obtained by Algorithm 4.3.

Figure 4.10d shows bandwidth obtained versus sampling interval with $C = 10$. Note that the algorithm guarantees the minimum bandwidth to the system. Here the bandwidth obtained from Algorithm 4.3 increases in an up down fashion. The randomness is due to the random channel availability and the increase in overall bandwidth is due to the selection of channels having high bandwidth and high switching. Figure 4.10c shows a decrease in the difference between the switching obtained by Algorithm 4.3 and the allowed switching, and is accounted for increase in bandwidth as shown in Fig. 4.10d.

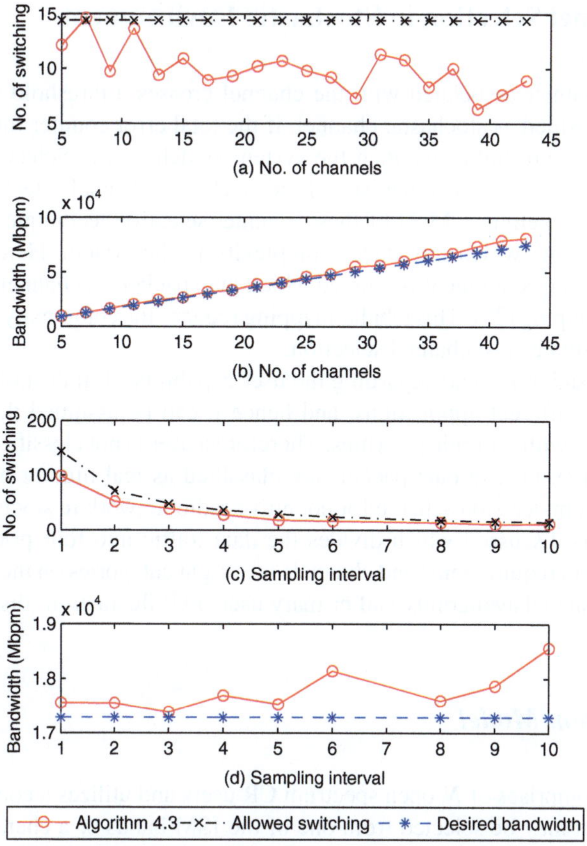

Fig. 4.10 Performance analysis of Algorithm 4.3 with respect to desired bandwidth and allowed switching

4.3.9 Ad-hoc Mode Operation

The channel selection techniques discussed in the previous section assumed the system model of Single Network in Infrastructure Mode. The discussed channel selection techniques can be modified to apply for networks operating in ad-hoc mode. To do so, additional protocol is required to achieve the goal of global channel availability matrix which is same for all the CR users in the ad-hoc network. If every CR user has same channel availability matrix, then the channel selection algorithm with desired constraints operating over all CR users will result in common channel selection, which can be utilized by an ad-hoc media access scheme for communication purposes.

4.4 Channel Selection in Stochastic Environment

If the error counter associated with the channel crosses a threshold (ξ), then that channel is classified as stochastic channel. If the total error counter for all channels crosses a given threshold (Ψ), then the system switches onto Selection on the go approach. The system may utilize any of the stochastic channel selection technique available in the literature. Most of these channel selection technique are based on statistical learning, which introduces complexity to the system. Here a stochastic channel selection technique is discussed which map packets to channels on the basis of a tabular mapping [28]. This tabular mapping reduces the complexity of the system as well as facilitates fast channel selection.

A generalization is made regarding the user capability of multi-tasking and running multiple different applications, and hence it can be assumed that a user can generate traffic with multiple priorities. Therefore a user is not classified as real time or non-real time, instead data packets are classified as real time or non-real time. Here, a spectrum decision scheme for cognitive radio network in stochastic channel environment is discussed which divides the data traffic into four priorities on the basis of the QoS requirements and channels into eight categories on the basis of their bandwidth, channel availability and primary user (PU) fluctuation (discussed later).

4.4.1 System Model

The system comprises of N open spectrum CR users and utilizes a common control channel which may be selected from one of the ISM bands or a channel with high availability. Each of the C channels utilized in the system has a bandwidth of B_i, $\forall i \in \{1, \ldots, C\}$. Each channel is associated with an ID and counter for its identification and load balancing purposes (discussed later) respectively. Therefore a total of C counters is maintained by each node in the network. Each CR user is in communication range with all other CR users and the transmission range of a PU is more than that of the CR user.

As discussed in Chap. 3, the user traffic is classified into four different priorities as shown in Table 3.1. It is assumed that the application layer is capable of tagging (marking) the packets associated with a particular traffic. The channel availability is modelled as two-state Markov chain as discussed in Sect. 3.4.1. It is assumed that each CR user statistically learns the channel transition probabilities from their previous sensing. Therefore each CR user knows the channel transition probabilities for all C channels.

4.4.2 Communication Segment

A communication segment (t, r) is defined as a pair of transmitting and receiving CR users for a session. A communication segment operates over a single channel and does not spawn over multiple channels during a session. Also a communication segment can carry only a single class (type/priority) of data and therefore has the priority depending upon the type of data traffic it is carrying. This section will use the word 'connection' interchangeably with the word 'communication-segment'.

4.4.3 Spectrum Decision and Mapping of Packets

The spectrum decision involves selection of channels by CR users depending upon the data packet needed to be sent. All C channels are classified into eight types on the basis of bandwidth, channel availability and PU fluctuation. The calculation of channel availability and fluctuation is discussed later. Table 4.2 shows the tagging of these channels. The term 'High' and 'Low' are subjective and may vary depending upon application and licensed channels. Voice packets need those channels which have higher availability and low fluctuation. An example would be if $\alpha = 0.9$ and $\beta = 0.9$; the channel is although available with probability of 0.5 but it alternates at each time slot, resulting in jittering at each step. Therefore PU fluctuation need to be taken care of by using fluctuation parameter. The fluctuation parameter δ is defined as:

$$\delta_i = \frac{\alpha_i + \beta_i}{2} \; \forall i \in C' \tag{4.7}$$

The fluctuation parameter $\delta \in [0, 1]$ tells how much the PUs' presence fluctuate (varies) on the given channel. A higher value of δ represents high fluctuation while

Table 4.2 Tagging and classification of primary licensed channels

Bandwidth	Channel availability	Primary user fluctuation	Tag
High	High	High	0
High	High	Low	1
High	Low	High	2
High	Low	Low	3
Low	High	High	4
Low	High	Low	5
Low	Low	High	6
Low	Low	Low	7

Table 4.3 Mapping of data packets to channel requirements

Access category	Bandwidth	Channel availability	Primary user fluctuation
Voice	High/Low	High	Low
Video	High	High	High/Low
Best effort	Low	High/Low	Low
	High	High/Low	High/low
Background	Low	High/Low	High/Low

a low value of δ represents low fluctuation. Fluctuation parameter only represent the PU fluctuation and does not represents the actual availability of the channel.

Therefore high availability of the channel and minimal fluctuation of the PU on that channel is desired for the system supporting real time traffic like voice. A quick analysis of Eqs. 3.2 and 4.7 indicates that these criteria can be met with low value of β and equal value of α and β. The low value of β decreases the fluctuation of the PU on the channel while an equal value of α and β increases the channel availability.

Once the quantity 'High' and 'Low' are associated with bandwidth, availability of all the channels along with the PUs fluctuation on it, the CSMU can define mapping function for mapping packets to channels. The mapping function is defined in Table 4.3. The table shows that the voice packets can choose channels with low bandwidth, as the bandwidth of voice packet is small. The guarantee is made in terms of channel availability and PU fluctuation on the channel. Similarly, video packets can choose a channel with high PU fluctuation as videos can be buffered if needed. Background traffic can only choose channels with low bandwidth, whereas Best Effort data considerably overlaps with other data priorities. The reason is that if video or voice traffic is not present then best-effort traffic can use the channel. Each traffic starts with selecting low bandwidth channel in its class i.e. a voice and best-effort traffic will first try to choose a low bandwidth channel and if not available then move to next higher bandwidth channel.

4.4.4 *Load Balancing*

Since there are eight categories of channel as shown in Table 4.2, the numbers of channel in a single category is represented by γ_i, $\forall i \in C'$. Therefore $\sum_{i=0}^{7} \gamma_i = C$. It may be possible that some category may not have any channels in them. Let ξ_k, $\forall k \in \{0..7\}$ represent the set of channels IDs in a given category. The purpose of the load balancing system is to uniformly distribute the communication segments within the channels of a given category. Algorithm 4.4 provides the method of load balancing to the system while doing the spectrum decision for a communication segment. The index for ξ is dropped because the algorithm represents the load balancing within a category. The variable array **cntr** maintains the list of counter for all the channels, and it is used to count the activities on a channel. The counter value for each channel is set to zero during the initialization of the system. The sensing report unit of the framework updates the value of the counter based on the external activity (PU or any other CR user) whereas the counter variable is also updated inside the algorithm block to count for the internal activity. The higher the value of the counter for a channel, the higher the activities it has got on it. The goal of the algorithm is to select a channel within a given category of channels such that it has minimal activity on it. If a channel with minimal number of activity is selected and used, a report is generated with the help of success/failure unit of the framework (lines 6–8). If the success reported is negative, then that channel is removed from the category for the current session and minimal activity channel is selected from the rest of the channels (lines 9–12). If all channels are removed then the algorithm starts from the beginning. The counter is reset to 0 once it reaches a threshold. The threshold can vary according to the application.

4.4.5 Performance Analysis

The terms 'High' and 'Low' for bandwidth of a channel, channel availability and PU fluctuation on a given channel are defined in order to obtain the performance analysis. Any channel having bandwidth greater than or equal to average bandwidth is called as 'High' bandwidth or otherwise 'Low' bandwidth. The average bandwidth is calculated using Eq. 4.8.

$$avg\_band = \frac{\sum_{i=1}^{C} B_i}{C} \tag{4.8}$$

Algorithm 4.4: Algorithm for load balancing in a given category of channels

1 **Initialize** :$\psi = \xi$;
2 *Label* :
3 **if** $\psi = \{\}$ **then**
4     $\psi = \xi$;
5 **end**
6 $ch\_id = argmin_{i \in \psi}(\mathbf{cntr}_i)$;
7 $\mathbf{cntr}_i = \mathbf{cntr}_i + 1$;
8 $report = success\_report(ch\_id)$;
9 **if** $report = 0$ **then**
10     $\psi = \psi - \{ch\_id\}$;
11   goto Label;
12 **end**

Therefore,

$$Bandwidth_i = \begin{cases} High, & \text{If } Bandwidth_i \geq avg\_band \\ Low, & \text{otherwise} \end{cases} \tag{4.9}$$

Channel availability and PU fluctuation is similarly assigned as 'High' and 'Low' as follows.

$$avg\_ch\_av = \frac{\sum_{i=1}^{C} \Pi_i}{C} \tag{4.10}$$

$$ChannelAvailability_i = \begin{cases} High, & \text{If } \Pi_i \geq avg\_ch\_av \\ Low, & \text{otherwise} \end{cases} \tag{4.11}$$

$$avg\_pu\_fl = \frac{\sum_{i=1}^{C} \delta_i}{C} \tag{4.12}$$

$$PuFluctuation_i = \begin{cases} High, & \text{If } \delta_i \geq avg\_pu\_fl \\ Low, & \text{otherwise} \end{cases} \tag{4.13}$$

The discussed channel selection scheme is implemented on top of the MAC protocol discussed in Chap. 5. The discussed scheme is evaluated using discrete event simulation in Matlab which simulates the network up to frame level. The performance of the discussed framework is evaluated in terms of connection fail ratio, number of connections failed and number of channels used. Connection fail ratio is calculated by

$$\text{Connection fail ratio} = \frac{\text{number of failed connections}}{\text{number of connections initiated}}$$

The parameters α and β are assumed to be uniformly distributed over the range [0, 1] i.e. the transition probabilities of the channels are uniformly distributed over the range [0, 1]. The bandwidth of the channel is modelled to be linear with the number of channels used, i.e. the minimum bandwidth of a channel is given a value of 50 Kbps and the maximum bandwidth of the channel is $50 * C$ Kbps, where C is the maximum number of channels used for the simulation. Therefore the bandwidth is distributed linearly from channel 1 to channel C as 50 to $50 * C$ Kbps.

Figure 4.11 shows the graph of connection fail ratio versus the total number of channels used the system. The number of voice communication segment, video communication segment, best-effort communication segment and background traffic communication segment used are 2, 5, 30 and 13 respectively giving the percentage ratio of 4, 10, 60 and 26 % for voice, video, best-effort and background traffic communication segments respectively. The number of channels is varied from 10 to 40. The connection fail ratio decreases with increase in number of channels for all types of communication segments. Since the number of voice traffic is small and the number of channels is adequate to satisfy the requirement, the connection fail ratio for

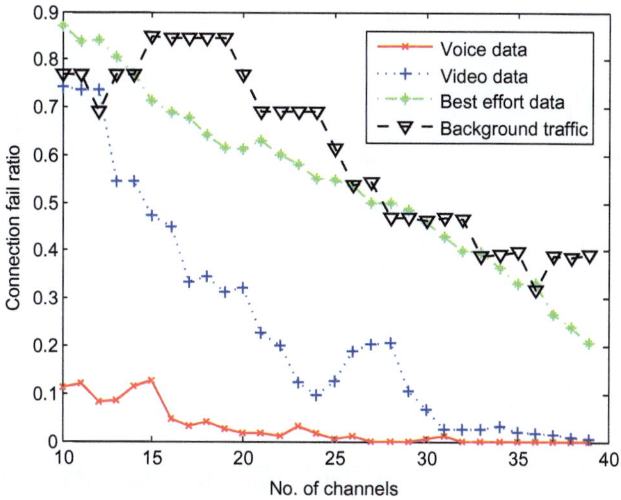

Fig. 4.11 Connection fail ratio versus number of channels

voice traffic is smallest. Background traffic shows higher blockage at start because most of the channels are used by the best-effort traffic (since there is overlapping of the channels for these two traffic types in mapping Table 4.3).

Figure 4.12 shows the graph of number of failed connections versus total number of communication segments for $C = 50$. The total number of communication segments comprises 4, 10, 60 and 26 % of voice, video, best-effort and background traffic communication segments respectively. From the graph it is clear that the num-

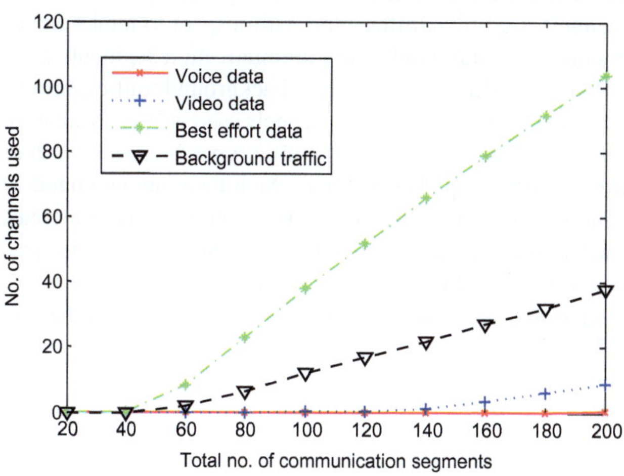

Fig. 4.12 Number of connections failed versus total number of communication segments

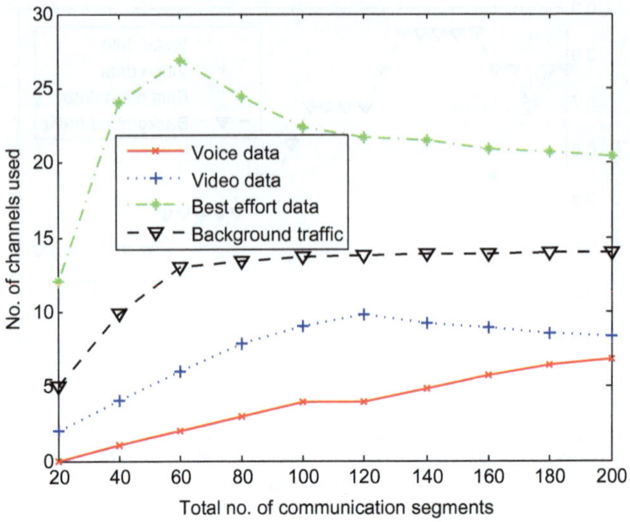

Fig. 4.13 Number of channels used versus total number of communication segments

ber of connections will fail as the number of connections (communication segments) increases. A steep increase can be seen when total number of communication segments equals 60, which can be explained by the fact that channels become busy after this point leading to failure in new connections. There is steep increase in the best-effort segment losses because of the large share of the total segments and some of the channels mapped to this traffic type is used by higher order traffic too (Table 4.3).

Figure 4.13 shows the load balancing in the system by taking in to account the number of channels used by a traffic type with respect to total number of communication segments. The total number of communication segments comprises 4, 10, 60 and 26 % of voice, video, best-effort and background traffic communication segments respectively. The total number of channels used is 50. It is clear from the graph that the number of channels used for all traffic types increases till total number of communication segments equals 60. After which there are no channels to fulfil the upcoming requests. As the number of communication segment increases, the number of voice and video data communication segments also increase leading to more use of the channels mapped by voice and video data. Thus there is decrease in the channel use by the web traffic data which were overlapped with the voice and video data (Table 4.3).

4.5 Discussion

The performance of a network depends upon the bandwidth and quality of the wireless channel in which it operates. A channel can be characterized on the basis of its availability; and on the basis of observation, channels may have deterministic/non-deterministic usage pattern. Channels like TV and FM bands can be utilized for the purpose of modelling deterministic behavior of licensed users. If channels do not have any deterministic usage pattern, then stochastic channel selection techniques (Selection on the go approach) are utilized.

In this chapter we have discussed a reconfigurable channel selection technique which initially starts with deterministic channel selection techniques for channels with deterministic usage behavior. If channels do not follow a deterministic behavior, the increase in error counter forces the system to utilize another discussed channel selection technique for stochastic environment. Three different algorithms have been discussed in deterministic channel environment targeting different aspects. These algorithms solve the problem of minimal channel switching and high throughput. Channel switching is reduced to mitigate the effects of packet losses, delays and jitter caused by channel switching. One of the discussed algorithm provides an intermediate result between minimal channel switching and maximum throughput, solving the high throughput minimal channel switching problem. In stochastic channel environment, the given technique divides channel into into eight different categories based on bandwidth, channel availability and PU fluctuation parameter. A mapping is provided from the data traffic priority to the channel type for the purpose of channel selection, along with a load balancing algorithm to balance the traffic on different channels.

Performance analysis of channel selection techniques in deterministic environment is done with three other channel selection techniques. Simulation result shows that the performance of the discussed algorithms based on the given system model approach provides better performance of the overall system. In addition the given model can provide a trade-off between minimal channel switching and maximum bandwidth. Similarly, in stochastic environment, the performance analysis shows that the channel selection can be utilized for obtaining different priorities with effective load distribution. Although this chapter covered the schemes for spectrum selection in deterministic and stochastic channel environment, it misses the important parameter of energy consumption.

References

1. Mishra, V., Tong, L. C., & Syin, C. (2013). Deterministic time pattern based channel selection in cognitive radio network. In *The International Conference on Information Networking 2013 (ICOIN)* (pp. 169–174).
2. Erpek, T., Lofquist, M., & Patton, K. (2007). Spectrum occupancy measurements loring commerce centre limestone, Maine September 18–20, 2007. Shared Spectrum Company Report.

3. Erpek, T., Steadman, K., & Jones, D. (2007). Dublin Ireland spectrum occupancy measurements collected on April 16–18, 2007. Shared Spectrum Company Report.
4. McHenry, M. A., McCloskey, D., Roberson, D., & MacDonald, J. T. (2005). Spectrum occupancy measurements Chicago, Illinois November 16–18, 2005. Shared Spectrum Company Report.
5. McHenry, M. A., & Steadman, K. (2005). Spectrum occupancy measurements location 1 of 6: Riverbend park, Great Falls, Virginia April 7, 2004. Shared Spectrum Company Report.
6. McHenry, M. A., & Steadman, K. (2005). Spectrum occupancy measurements location 2 of 6: Tyson's square center, Vienna, Virginia April 9, 2004. Shared Spectrum Company Report.
7. McHenry, M. A., & Chunduri, S. (2005). Spectrum occupancy measurements location 3 of 6: National science foundation building roof April 16, 2004. Shared Spectrum Company Report, Revision 2.
8. McHenry, M. A., & McCloskey, D. (2005). Spectrum occupancy measurements location 4 of 6: Republican national convention, New York City, New York August 30, 2004–September 3, 2004. Shared Spectrum Company Report, Revision 2.
9. McHenry, M. A., & Steadman, K. (2005). Spectrum occupancy measurements location 5 of 6: National radio astronomy observatory (NRAO) Green Bank, West Virginia October 10–11, 2004. Shared Spectrum Company Report, Revision 3.
10. McHenry, M. A., & Steadman, K. (2005). Spectrum occupancy measurements: Shared spectrum building roof, Vienna, Virginia, December 15–16, 2004. Shared Spectrum Company Report.
11. Valenta, V., Marsalek, R., Baudoin, G., Villegas, M., Suarez, M., & Robert, F. (2010). Survey on spectrum utilization in europe: Measurements, analyses and observations. In *Proceedings of the Fifth International Conference on Cognitive Radio Oriented Wireless Networks Communications (CROWNCOM)* (pp. 1–5).
12. Patil, K., Skouby, K., Chandra, A., & Prasad, R. (2011). Spectrum occupancy statistics in the context of cognitive radio. In *14th International Symposium on Wireless Personal Multimedia Communications (WPMC)* (pp. 1–5).
13. Sanders, F. (1998). Broadband spectrum surveys in Denver, CO, San Diego, CA, and Los Angeles, CA: Methodology, analysis, and comparative results. In *IEEE International Symposium on Electromagnetic Compatibility* (Vol. 2, pp. 988–993).
14. Chiang, R., Rowe, G., & Sowerby, K. (2007). A quantitative analysis of spectral occupancy measurements for cognitive radio. In *IEEE 65th Vehicular Technology Conference, 2007. VTC2007-Spring* (pp. 3016–3020).
15. Wellens, M., Wu, J., & Mahonen, P. (2007). Evaluation of spectrum occupancy in indoor and outdoor scenario in the context of cognitive radio. In *2nd International Conference on Cognitive Radio Oriented Wireless Networks and Communications, 2007. CROWNCOM* (pp. 420–427).
16. Yin, S., Chen, D., Zhang, Q., Liu, M., & Li, S. (2012). Mining spectrum usage data: A large-scale spectrum measurement study. *IEEE Transactions on Mobile Computing, 11*, 1033–1046.
17. Lopez-Benitez, M., Umbert, A., & Casadevall, F. (2009). Evaluation of spectrum occupancy in spain for cognitive radio applications. In *IEEE 69th Vehicular Technology Conference, 2009. VTC Spring* (pp. 1–5).
18. Bao, V. N. Q., Cuong, L. Q., Phu, L. Q., Thuan, T. D., Quy, N. T., & Trung, L. M. (2011). Vietnam spectrum occupancy measurements and analysis for cognitive radio applications. In *2011 International Conference on Advanced Technologies for Communications (ATC)* (pp. 135–143).
19. Islam, M., Koh, C., Oh, S. W., Qing, X., Lai, Y., Wang, C., et al. (2008). Spectrum survey in Singapore: Occupancy measurements and analyses. In *3rd International Conference on Cognitive Radio Oriented Wireless Networks and Communications, 2008. CROWNCOM* (pp. 1–7).
20. Cordeiro, C., Challapali, K., Birru, D., & Sai Shankar, N. (2005). IEEE 802.22: The first worldwide wireless standard based on cognitive radios. In *First IEEE International Symposium on New Frontiers in Dynamic Spectrum Access Networks, 2005. DySPAN* (pp. 328–337).
21. Mishra, V., Lau, C. T., & Chan, S. (2013). Reconfigurable channel selection technique for cognitive radio network with heterogeneous primary bands. *IEEE Transactions on Vehicular Technology, 62*, 3942–3952.

22. Song, Y., & Xie, J. (2010). Common hopping based proactive spectrum handoff in cognitive radio ad hoc networks. In *IEEE Global Telecommunications Conference (GLOBECOM 2010)* (pp. 1–5).
23. Lee, W.-Y., & Akyldiz, I. (2011). A spectrum decision framework for cognitive radio networks. *IEEE Transactions on Mobile Computing*, *10*, 161–174.
24. Garey, M. R., & Johnson, D. S. (1990). *Computers and Intractability; A Guide to the Theory of NP-Completeness*. New York: W. H. Freeman & Co.
25. Working group on wireless regional area networks, IEEE Std. 802.22, 2011. http://www.ieee802.org/22/.
26. Song, Y., Fang, Y., & Zhang, Y. (2007). Stochastic channel selection in cognitive radio networks. In *IEEE Global Telecommunications Conference, 2007. GLOBECOM'07* (pp. 4878–4882).
27. Bennai, M., Sydor, J., & Rahman, M. (2010). Automatic channel selection for cognitive radio systems. In *IEEE 21st International Symposium on Personal Indoor and Mobile Radio Communications (PIMRC)* (pp. 1831–1835).
28. Mishra, V., Tong, L. C., & Syin, C. (2012). Qos based spectrum decision framework for cognitive radio networks. In *18th IEEE International Conference on Networks (ICON)* (pp. 18–23).

Chapter 5
Case Study: Media Access in CRN Framework

Based on the survey done in Sect. 1.7 and Chap. 2, the media access control protocols are classified into two broad categories: *random* and *time-slotted* media access schemes, each having its pros and cons. Recently a new class of access scheme, known as *hybrid* schemes has been developed with better performance than that of the random and time-slotted access schemes. Most of these scheme are network-model dependent, and therefore a change in their operation from ad-hoc to infrastructure or vice-versa is impractical. In a multi-channel environment, support for broadcast and multicast is tedious in nature and requires implementation of flooding operation.

In this section, a media access scheme is discussed which is hybrid in nature and provides unique features of multicast and broadcast without network flooding [1]. The discussed media access scheme allows contention of nodes to be distributed across channels, therefore mitigating the problems of channel saturation. The scheme also allows nodes to prioritize packets for the purpose of QoS provisioning while facilitating power saving mode for energy conservation.

As discussed in Sect. 3.2, the units responsible for the design of QoS provisioning media access scheme are highlighted in Fig. 5.1. The mac protocol unit (MPU) receives the list of channels from channel selection and management unit (CSMU). It also receives the information about the QoS requirement on long run statistical behavior from the QoS control unit. Based on the information received the MPU generates a MAC instance. The MAC instance on the basis of QoS requirement of individual packet (access category) defines the individual packet behavior.

5.1 System Model

The system comprises of N cognitive radio user operating in an ad-hoc environment comprising of C channels. A channel from one of the ISM bands or a channel with high availability is selected as a rendezvous channel (RC). RC is utilized for the

V. Mishra et al., *QoS and Energy Management in Cognitive Radio Network*,
Signals and Communication Technology, DOI 10.1007/978-3-319-45860-1_5

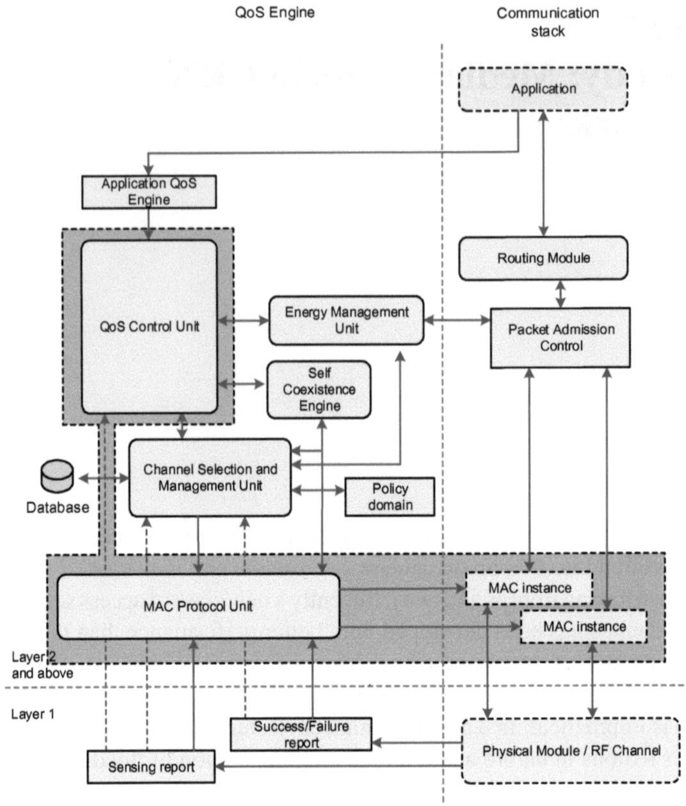

Fig. 5.1 Units in the CRN framework responsible for QoS provisioning media access scheme

purpose of network information interchange. The hardware complexity of CR users is equivalent to that of the half-duplex transceiver with sensing capability.

5.1.1 Traffic Type

In modern world, conventional wireless network has to support multitude of different data types due to increase in different types of applications. In a broad sense, data traffic can be classified into four categories as discussed in Sect. 3.1.1 and shown in Table 3.1. Some applications based on voice or video services require delay bounded access, which require special handling of such packets.

As discussed in Sect. 3.1.1, the resource access delay suffered by a packet in the network is governed by the access scheme utilized by the scheme. An access scheme utilizes packet admission control through the introduction of different arbitration spaces for different packets. Discussion on arbitration space is provided in Sect. 5.2.1.

Apart from arbitration space, QoS is governed by the reservation of bandwidth for a particular traffic. This ensures that the traffic is provided with enough bandwidth to support the level of QoS desired.

5.1.2 Channel Classification

The discussed scheme utilizes multiple channel for its operation, where different channels serves different purposes. The channels utilized in the system are classified into two types: *Data* and *Rendezvous* channels. Users in the network perform exchange of data packets over data channels while the exchange of control information is done over rendezvous channel (RC). The discussed media access scheme supports two types of data transfer: *contention* based and *reservation* based data transfer. The support for dual data transfer types requires usage of two different access etiquettes. Therefore data channels are further classified into *Contention based channels* and *Reservation based channels*. Contention based channels (CBC) follow the etiquette of contention based data transfer. Nodes which are on CBC perform contention whenever a data transfer is required. Reservation based channels (RBC) follow the etiquette of data transfer based on reservation. For the purpose of reservation, nodes perform contention ahead of real data transfer on RBC. Therefore this type of channel has two separate data period: contention (denoted by ad-hoc transmission indication message (ATIM) period) and reserved data period. A reserved data period is further sub-divided into data slots; each of which represents a data packet transfer. During contention, nodes contend to reserve data slots, and during reserved data period, scheduled nodes perform communication. Reservation based channels allow implementation of higher priority stream based data services (for example voice and video).

The system comprises of non-hopping nodes where nodes reside on a channel as long as channel ceases to be free or nodes want to switch to another channel for some reason. From a node's perspective, channels are classified as *Home channel* and *Foreign channel*. Home channel is a channel in which a node chooses to reside while all other channels from this node's perspective are termed as foreign channels. A node belonging to home channel is termed as *Home node* and all other nodes are termed as *Foreign nodes*.

5.1.3 Quiet Period Distribution

As discussed in Sect. 3.4.2, quiet period on a channel is a time duration during which all CR users on that channel stop their communication so that effective spectrum sensing can be performed. Quiet periods are designed to be periodic in nature, so that nodes can detect an incumbent activity during the start of the current/next frame. Sensing can be classified into two types: *in-band* and *out-of-band* sensing. In-band sensing refers to sensing of home channel by home nodes while out-of-band sensing refers to sensing done by nodes on foreign channels. To improve the overall throughput of the system, the quiet periods on all channels are distributed such that they do not overlap. This allows in-band and out-of-band sensing to be performed without any throughput loss.

5.2 Hybrid Media Access Scheme

The design of the discussed MAC scheme with two reservation based channels, two contention based channels and a rendezvous channel is shown in Fig. 5.2. Each channel in the system is divided into fixed length time frames known as superframes. The superframe of all data channels are overlapped and synchronized with each other. Rendezvous channel is synchronized with data channels with a forward offset value of size of rendezvous channel beacon (RCB). Therefore, the start of superframe of data channels are overlapped and synchronized with the end of the RCB of rendezvous channel.

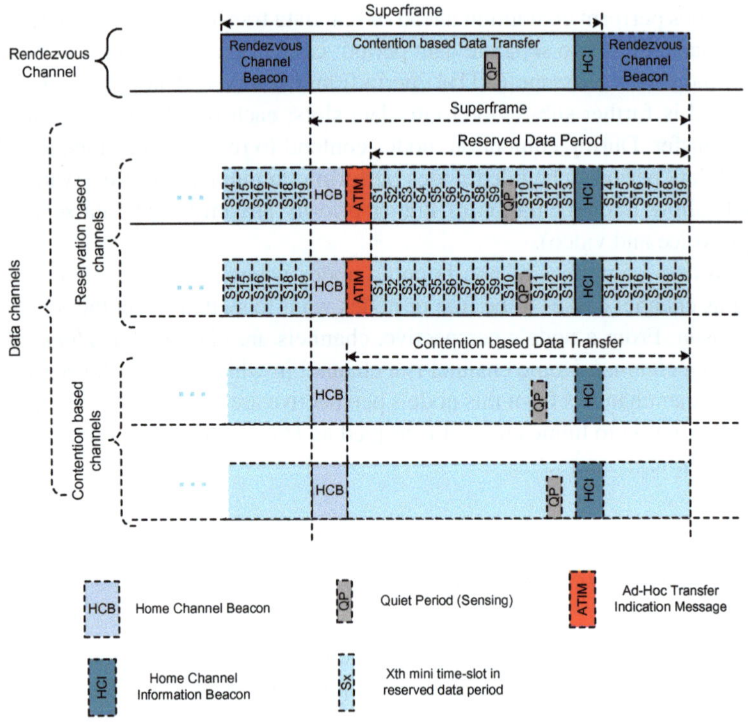

Fig. 5.2 MAC frame structure

The superframe of all channels is composed of different intervals designated for specific purposes. Common intervals among all superframes are quiet periods (QP) and home channel information beacon (HCI). During QP, all nodes on that channel stop their transmission and perform *in-band* sensing. During this period, any foreign node (i.e. node on some different channel than that of the current QP) can perform *out-of-band* sensing on this channel. During HCI, all nodes in their respective home channels compete to select a representative node (RN). A representative node is chosen periodically from each channel which broadcasts information about its home channel onto rendezvous channel. The information consists of number of nodes on home channel, quiet period distribution and channel characteristics. Apart from QP and HCI, rendezvous channel comprises of rendezvous channel beacon (RCB) and contention based data transfer period. A RCB has a time slotted design, with total number of slots designated as t. During RCB, representative nodes (RN) switch to rendezvous channel and share information about their home channel. Each slot in RCB is designated to each RN, therefore $t = C + 2$; where 2 additional slots are required for new nodes joining the network.

Apart from QP and HCI, the superframe of a data channel comprises of home channel beacon (HCB) and data transfer period. During HCB, the RN returns from rendezvous channel to its home channel, and broadcasts the information it learned from other representative nodes in the network. A HCB on a channel has a time slotted architecture, where number of time-slots equals the number of nodes on that channel +1. An extra time slot is provided for a new node joining the network on that channel. Depending upon data channel type, the data transfer can be classified into two types: *reservation* and *contention* based data transfer. A reservation based channel (RBC) employs reservation based data transfer service, while a contention based channel (CBC) employs contention based data transfer service. RBC consists of ATIM (ad-hoc transfer indication message) period and reserve data slots. During ATIM period, communicating node performs contention, and reserve data slots ahead of actual data transmission. A three way handshake is performed during the reservation process. The reservation of a resource block before actual communication, allows implementation of higher priority service. The design of contiguous resource blocks allows provisioning of stream based services like voice or video which require contiguous stream of data transfer [2]. On the other hand contention data period on CBC requires nodes to perform a 4 way handshake similar to optional RTS/CTS mechanism of 802.11 [3]. Therefore nodes contend whenever they require to perform a data transfer during the contention data period. Contention based data transfer service is also followed by nodes on rendezvous channel.

5.2.1 Arbitration Interframe Spaces

For the purpose of packet prioritization, different arbitration interframe spaces are utilized. Arbitration interframe space is the time duration which a node has to wait before transmitting a packet. Arbitration in this discussed scheme is different from

Table 5.1 Interframe spaces for different packet types

Interframe space	RTS packet type/Other packet type	IFS_{min}	IFS_{mxo}
SIFS	Voice/CTS/ACK/DATA	IFS_{smin}	IFS_{smax}
MIFS	Video	IFS_{mmin}	IFS_{mmax}
IIFS	Best effort	IFS_{imin}	IFS_{imax}
LIFS	Background data	IFS_{lmin}	IFS_{lmax}

the traditional arbitration as in enhanced distributed channel access (EDCA) [4], where a fixed value is utilized and the value of contention window is changed on each retransmission. Here arbitration is defined as a random duration in $[IFS_{min}, IFS_{max}]$. Here $IFS_{max} = 2^r.IFS_{mxo}$, where r is the number of retransmission performed for this packet, and IFS_{mxo} is the initial maximal boundary for arbitration value selection. Each packet before transmission is assigned a counter which is initialized with the value of arbitration space. The counter is decremented whenever the channel is sensed to be free. If the counter reaches zero, the packet is transmitted. If transmission is unsuccessful, then the value of r is incremented and the packet is scheduled for retransmission. If the packet is not deliverable even after Γ number of retransmission, it is dropped. The value Γ is called as retry limit. Different limits of arbitration interframe space for different types of packets are shown in Table 5.1, in which S/M/I/L-IFS implies short, medium, intermediate and large interframe space respectively. Therefore an RTS packet of video has to wait MIFS duration before transmission. Similarly ACK packet of any data type has to wait for SIFS duration before transmission.

5.3 Initialization

The cognitive radio network initializes by selecting a rendezvous channel. A node in the network scans for a RCB while performing sensing on all *defined* channels. If a node does not detect any RCB, then it starts its own RCB by first selecting an appropriate channel as a rendezvous channel (RC). An auxiliary RC may also be selected as a backup, just in case main RC ceases to be free. If a node detects an existing RCB, it joins through by doing transmission on the extra time-slots of RCB. Based on the information it learned from RCB, it may join a data channel by following the same procedure on HCB. A node having high probability of performing a high priority data transmission will select reservation based channel. If there does not exist a data channel, the node may start its own beacon for the data channel. The node may classify a data channel as RBC or CBC depending upon channel availability. A channel having high availability is considered for the candidate of RBC and vice-versa.

5.4 Operation

Since two different types of data channels are utilized in the access scheme, two different etiquettes of channel access are also utilized. The nodes follow the etiquette of the channel in which they perform communication. The communication may take place on either of the home channels of communicating nodes. If the desired characteristics of communication media is not supported by either home channels of the communicating nodes, then the communication may take place on a foreign channel. In this case the etiquettes of the foreign channel will be utilized for communication purposes. Depending upon the channel characteristic and location (home channel) of sender and receiver, three different types of data transfer are classified:

5.4.1 Data Transfer on Reservation Based Channels

On a reservation based channel, nodes contend to reserve data slots during ATIM period. For contending, nodes employ *three* way handshake to reserve data slots ahead of actual communication. A node initiates this by sending first request-to-send (RTS) packet. The RTS packet of sender consists of a list identifying available data time-slots onto which the sender intends to communicate. The receiver replies with clear-to-send (CTS) packet containing the list of data time-slots on which it agrees to receive the data. The sender confirms this list by sending an acknowledgement (ACK). All other nodes know which *slots* are reserved for what *type* of data transfer. A high priority data transfer like voice or video can preempt the reserved data slots from scheduled low priority data transfer service; and therefore high priority service can schedule its own transmission in those reserved slots as if they are not reserved. During the reserved data period, scheduled nodes perform data communication.

5.4.2 Data Transfer on Contention Based Channels

Data transfer on a contention based channel follows the traditional optional 4 way handshake RTS/CTS mechanism of 802.11 [5]. A sender node initiates the transmission by sending an RTS packet containing the duration of the transmission in a field called as network allocation vector (NAV). The receiver replies with CTS containing NAV, to identify that it is ready to receive the data. The sender then performs transmission of DATA packets. After the reception of DATA packet, the receiver replies with an acknowledgement (ACK) packet.

5.4.3 *Data Transfer on a Foreign Channel*

If a sender has a packet for a receiver which does not lie on the home channel of the sender, the sender moves to the home channel of receiver after HCB. The sender then performs communication with the receiver using the channel etiquettes of the receiver. Once the communication is finished or HCI of the home channel of the sender starts, the sender moves back to its own home channel. The procedure is repeated until the receiver node decides to move onto the home channel of the sender.

If the desired characteristic of the channel is not supported by the home channel of receiver, then data transmission is performed on a foreign channel with the desired characteristics. The sender moves to the home channel of the receiver and sends a MOV packet to the receiver indicating the channel into which they should switch onto. If the receiver agrees, it replies with an acknowledgement MOV-ACK and immediately switches to the indicated channel. The sender after receiving the acknowledgement also switches to the indicated channel. Both sender and receiver node waits for LIFS duration on the indicated channel to overhear any current communication and then follow their etiquettes. Both sender and receiver move back to their home channel when the communication is finished or HCI of the home channels of sender/receiver starts.

5.5 Power Saving Mode Operation

The discussed MAC scheme allows different power saving mode operation for nodes depending upon their home channel. Nodes are always active during the period of beacon exchange (RCB, HCB, HCI) on their home channel. Here active duration refers to the time duration for transmit, receive or sensing operation.

On a reservation based channel, nodes are always active during HCB, ATIM, QP and HCI period. A node learns about the usage of channel (data slots) during ATIM period, and therefore schedules to enter into doze (idle) mode if it is not performing any communication. The node wakes up when it has its turn for sending or receiving data on reserved data time-slots. Figure 5.3 shows 3 nodes A, B and C operating on a reservation based channel. During ATIM period, nodes A and B have scheduled communication on data time-slots $S5 - S14$. At the end of ATIM, nodes A, B and C enter into doze mode. At the start of data slot $S5$, node A and B become active and perform communication. During QP, node A and B stop their transmission and perform sensing operation. During QP, node C wakes up and performs sensing operation. During HCI, all nodes exchange their beacon for the purpose of selection of a RN. Since node A and B are busy in performing data communication, some other node is chosen as RN. Once the communication is done, node A and B again enter into doze mode.

The power saving operation on contention based channel is followed similar to the power saving mode (PSM) of 802.11 [3]. All operating nodes on the channel

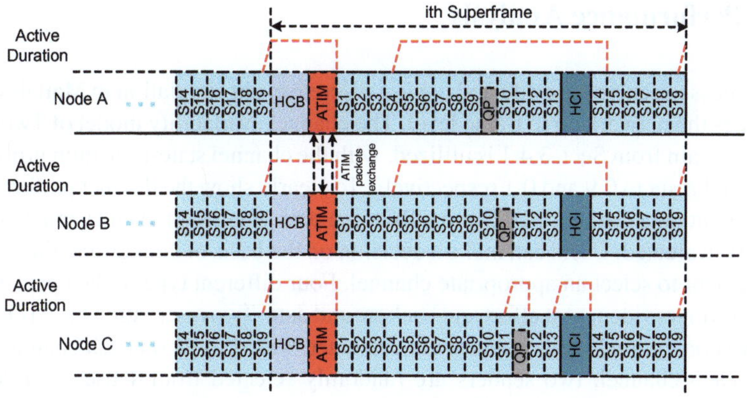

Fig. 5.3 Power saving mode operation on a reservation based channel of the MAC scheme

overhear the RTS and CTS packet of communicating nodes, which consists of a network allocation vector (NAV). Since NAV indicates the duration of transmission, all other nodes can calculate the time for which they should remain in doze mode. Therefore nodes sleep during an ongoing transmission of other nodes, and wake up when the transmission is finished. The value of NAV for RTS and CTS packet is calculated as:

$$NAV_{RTS} = t_{CTS} + t_{DATA} + t_{ACK}$$
$$NAV_{CTS} = t_{DATA} + t_{ACK}$$

where t_{pkt} is the time duration required for the transmission of a packet *pkt*.

5.6 Broadcast and Multicast Operation

Broadcast and multicast operation are required by many higher/lower layer protocols to implement certain functionality. Broadcast operation in multi-channel environment faces the problem of dissipation of a single packet over multiple channels and hence multiple channel hopping is required. The discussed scheme mitigates such a problem by providing a broadcast or multicast operation within 2 superframe. A node which has a packet to broadcast will select itself as a representative node and broadcast the packet in RCB of next superframe. The representative nodes from other channels on returning to their home channel from RCB will broadcast the packet in HCB, which allow all nodes on that channel to receive the broadcasted packet; since this follows on all channels, the broadcasted packet are received by all the nodes in the network. Multicast operation follows similar to the broadcast operation, except that a masking operation is required, which depends upon multicast application.

5.7 Performance Analysis

The discussed scheme is evaluated using a discrete event simulation in Matlab which simulates the network up to frame level. The channel availability model of Two-state Markov chain from Sect. 3.4.1 is utilized, with the channel state-transition probabilities, α and β set to 0.9 and 0.1 respectively. To clearly show the distinct performance on different data channel (RBC and CBC), the performance of each channel is individually evaluated. Users can make a decision on the basis of latency and throughput requirement to select an appropriate channel. Four different type of data packets are utilized: voice, video, best effort and background data. Two reservation based channel and two contention based channels are evaluated with 4 users on each channel.

On each channel, two senders are randomly selected from 4 users, while rest two as destination. The arrival rate of data packets on these users is modeled as Poisson arrival with the percentage distribution of aggregate arrival as voice, video, best effort and background data packets set as 10, 20, 60 and 10 % respectively. The parameters used for the simulation are shown in Table 5.2. To simplify the analysis, bandwidth of each channels is assumed to be equal and the size of RTS, CTS and ACK packets used on RBC and CBC are assumed to be equal. The data packet size of all 4 different data types are also assumed to be equal. To capture the effect of stream based transfer on RBC, a transmission of 3 voice packets requires only a single RTS/CTS/ACK handshake exchange during ATIM duration; similarly a transmission of 4 video packets requires only a single RTS/CTS/ACK handshake exchange during ATIM duration. While all other transmissions require exchange of RTS/CTS/ACK for every data packet. Similarly on CBC, all transmissions require exchange of RTS/CTS/ACK for every data packet. The arrival rate for each user is increased from 50 packets per second to 330 packets per second with step size of 20. The simulation time is 100 s, and each data point in the graph is the average of 100 run.

In Fig. 5.4, analysis of average latency of individual (in terms of priority) packets is done. A clear comparison of Fig. 5.4a, b evaluates that, packets on RBC has a low and defined (calculable) latency in comparison to CBC. Moreover the latency for voice and video packets in RBC does not increase as they tend to increase in CBC with increase in arrival rate. A calculable latency on RBC is attributed to its reservation based design, which allows data slots to be reserved ahead of actual communication. The main drawback of RBC etiquette in comparison to CBC is that, it has a high latency for low level priority packets.

Figure 5.5 shows the probability of success of transmission of each originating packet. From Fig. 5.5a, b, it can be concluded that as the arrival rate increases, there is a decrease in success probability due to more number of collisions. Also the success ratio of RBC for higher priority packets (voice and video) is more in comparison to CBC. This can be attributed to pre-scheduled DATA packets in RBC, where as in CBC, large DATA packets result in backoff or collision of other packets. The success ratio of low priority (best effort and background) data packets is lower in RBC due

Table 5.2 Simulation parameters

Parameters	Value
Simulation time	100 s
Slot time	10 μs
Number of RBC	2
Number of CBC	2
Bandwidth of channels	2 Mbps
Superframe size	100 ms
RCB beacon size	15 ms
HCB beacon size	3 ms
HCI beacon size	2 ms
Quiet period size	2.5 ms
ATIM window size	35 ms
Number of reserved data slots	24
Reserved data time slot size	2.5 ms
RTS packet size	240 bits
CTS packet size	240 bits
ACK packet size	240 bits
DATA packet size	5000 bits
IFS_{smin}, IFS_{mmin}	1
IFS_{smin}, IFS_{mmin}	1
IFS_{imin}, IFS_{lmin}	3
IFS_{smax}	3
IFS_{mmax}	5
IFS_{imax}	7
IFS_{lmax}	10
Γ	5

to the preemptive nature of high priority packets to take away the communication segments (data slots) reserved for low priority packets.

Figure 5.6 shows the channel utilization of RBC and CBC. Here, the channel utilization is calculated on the basis of actual DATA packets transmitted. Therefore, the channel utilization does not include packets like RTS, CTS or ACK. At low arrival rate, RBC tends to perform better due to low collision in ATIM contention window. The lower collision is attributed to small size of control packets, where actual large DATA packets are communicated during reserved data slots. But after a threshold (aggregate number of arrivals per node = 170), CBC performs better in terms of channel utilization. This is attributed to restricted contention window size in RBC. The small ATIM contention window in RBC cannot accommodate more than a defined number of control packets, and hence actual communication is limited by it. Therefore, CBC achieves a higher channel utilization due to large contention based data transfer period.

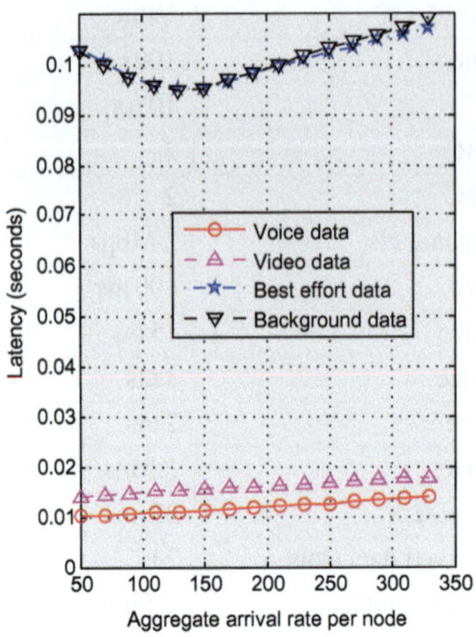

a: Reservation based channel

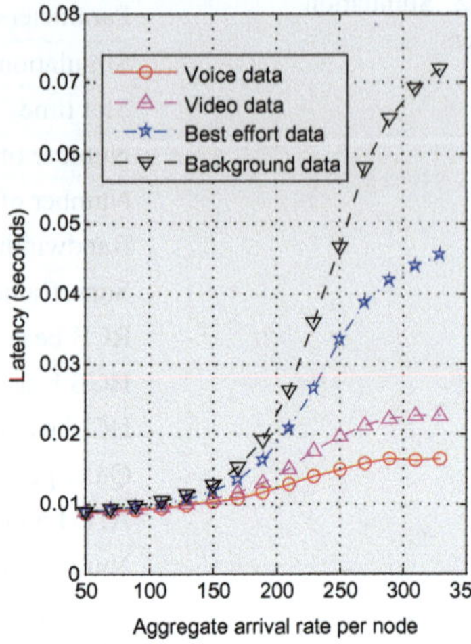

b: Contention based channel

Fig. 5.4 Latency of packets on different channels

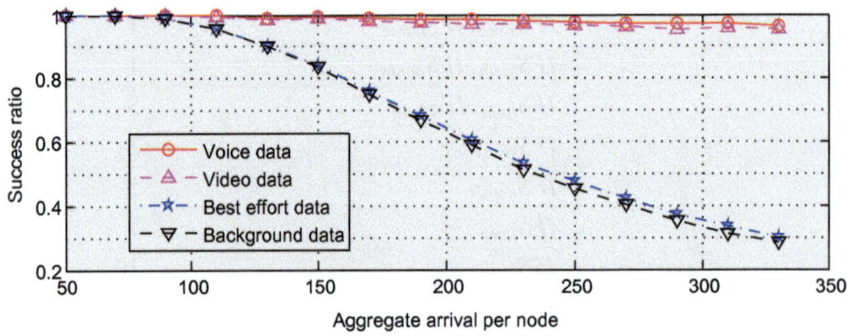

a: Reservation based channel

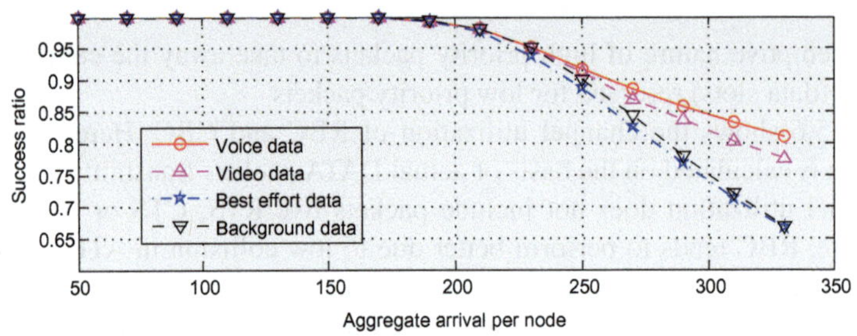

b: Contention based channel

Fig. 5.5 Success ratio of packets on different channels

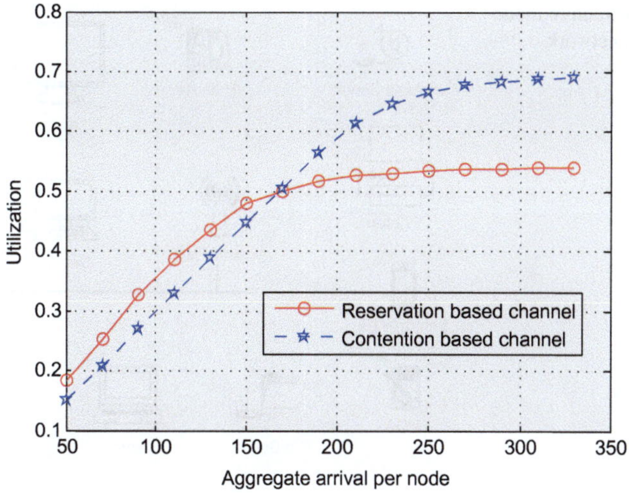

Fig. 5.6 Channel utilization

5.8 Infrastructure Mode Operation

The discussed scheme can also be employed in infrastructure mode by doing slight modification in middleware without any change on hardware requirements of users node. For operation in infrastructure mode, network employs an access point (AP) with the number of transceivers equal to the number of data channels employed. Data channels are formulated by AP with quiet periods distributed in a predefined sequence. The requirement of a rendezvous channel is removed and the AP acts as a *gateway* to all the nodes operating in the network. Access point acts as representative node (RN) on all data channels and the requirement of home channel information (HCI) beacon is also removed. Therefore all data channels (including RBC and CBC) are synchronized, and overlap with each other. During HCB, AP learns about the number of active nodes on a channel, channel characteristics and information regarding channel sensing from different nodes to perform cooperative sensing. Any node that wants to send a packet to another node, passes it through access point. Therefore user nodes do not hop for packet transmission. An AP also calculates and broadcasts average latency of packets on a given channel, so that a node can join/move to a given channel depending upon the observed latency. Nodes join/leave a channel through the help of AP; while AP also coordinates in assignment/removal of beacon slot for users node in HCB.

The infrastructure mode, operation of the discussed MAC scheme is shown in Fig. 5.7, where different equipments in home/office with different priority services operate with each other. An access point acts as a *gateway* and allows Internet connection through a *switch*. High priority services like IP phone and smartphone will operate with voice/video priority services. IP TV and media servers will operate with

Fig. 5.7 Infrastructure mode operation with network equipments

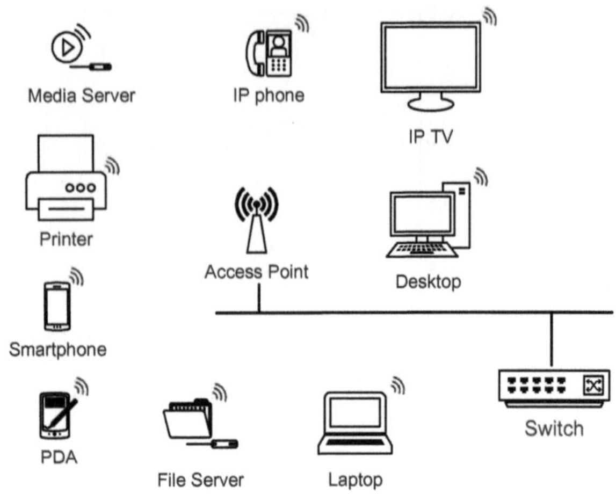

video priority but laptops and desktop will operate with hybrid priority (video and best effort) depending upon latency requirement, file servers and PDA will operate with best effort priority; while printers will operate with background priority.

In infrastructure mode, all nodes are connected to the AP, and hence are not hidden from it. Nodes may be hidden from each other, due to location and transmission power constraints. Since nodes perform three way or four way data exchange (depending upon channel type), the hidden terminal problem is solved.

5.9 Hidden Terminal Problem in Ad-hoc Mode

The discussed access scheme in ad-hoc mode is only defined for a network consisting of single-hop nodes. In multi-hop scenario, it may be possible that the information broadcasted by a rendezvous node (RN) is not be heard by some nodes on that channel due to the difference in location. Similarly, the information transmitted by some nodes on home channel beacon may not be heard by other home nodes. This is called hidden terminal problem, which results in nodes having inconsistent set of information. Multiple solutions exist to solve this problem. One solution is to increase the size of home beacon and ask every home node to repeat the information it learned from RN and other home nodes; therefore the information broadcasted is cumulative in nature. A node which is hidden to RN or to any other node, will overhear it and thus they all will have consistent set of information. Another solution is that all nodes generate its next-hop neighbor list. A next hop neighbor list of a *node* consists of list of neighbor nodes that the *node* learns from overhearing packets on the channel. This next hop neighbor list is matched with the information broadcast on HCB. If there are inconsistencies among the list of visible nodes on the channel, they are resolved.

5.10 Discussion

In a cognitive radio network, multiple cognitive nodes try to access sporadically varying resources (frequency spectrum) which results in access collision. To mitigate access collision, an access scheme is required which is followed by all the cognitive nodes operating in the cognitive radio network. The etiquettes of an access scheme is directly responsible for quality of service from latency perspective, while it is also responsible for quality of service from throughput perspective to an extent. The queuing delay which directly affects latency is controlled by the access scheme while throughput is directly controlled by the channel selection scheme and indirectly by the design of media access scheme.

In this chapter, different media access schemes for cognitive radio networks were discussed and a case study was performed on a QoS aware hybrid media access scheme framework for cognitive radio network. The scheme utilizes a hybrid of two different channel etiquettes: reservation based channel and contention based channel. These channels have different attributes and hence support different types of services. The discussed media access scheme utilizes power saving mode for doze mode operation while supporting major network operations like broadcast and multicast. An infrastructure based operation is also discussed which requires slight modification on middleware, while no changes are required on hardware of user node. Hidden terminal problems in infrastructure mode and ad-hoc mode are also discussed. Performance analysis of the discussed scheme shows that reservation based channels have low latency and higher transmission success probability for higher priority services. In contrast, the channel utilization of contention based channel achieves better performance than that of the reservation based channel for higher arrival rate. Even though a media access scheme mitigates access collisions within a network, it cannot directly mitigate access collision with another network.

References

1. Mishra, V., Tong, L.C., Syin, C. (2013) A qos provisioning mac protocol for cognitive radio network. In *2013 7th International Conference on Signal Processing and Communication Systems (ICSPCS)* (pp. 1–8).
2. Mishra, V., Tong, L.C., Chan, S., Mathew, J. Mac protocol for two level qos support in cognitive radio network. In *2011 International Symposium on Electronic System Design (ISED)* (pp. 296–301).
3. IEEE 802.11 working group, wireless lan medium access control (mac) and physical layer (phy) specifications. IEEE 802.11 Standard (2012).
4. Perez-Costa, X., & Camps-Mur, D. (2010). IEEE 802.11e qos and power saving features overview and analysis of combined performance [accepted from open call]. *IEEE Wireless Communication*, *17*, 88–96.
5. Sakurai, T., & Vu, H. (2007). Mac access delay of ieee 802.11 dcf. *IEEE Transactions on Wireless Communication*, *6*, 1702–1710.

Chapter 6
Case Study: Energy Management in CRN Framework

In Chap. 2, different spectrum selection strategies were discussed for the purpose of QoS provisioning in deterministic and stochastic channel environment. In this chapter, an energy aware spectrum decision scheme for QoS provisioning CR framework (ESDF-CR) is discussed in which a user selects a channel which reduces the power consumption of transmission [1]. The scheme kicks in when the remaining energy of the user falls below a given threshold, so as to allow extended operation of the system with graceful degradation in QoS. The selected channel has lower bandwidth and thus has limited throughput capacity for transmission which results in queuing of packets. A packet admission control scheme is designed so as to allow high priority packets to pass through the selected channel while blocking the lower priority packets. Thus a graceful degradation of QoS can be maintained by the system.

The technologies/modules responsible for this energy aware spectrum decision are highlighted in the discussed QoS provisioning framework as shown in Fig. 6.1. The energy management unit (EMU) provides the information and constraints regarding the remaining power of the system, whether it is running on AC power or battery. When the remaining power of the system fall below a certain threshold, the EMU asks for energy aware spectrum decision through channel selection and management unit (CSMU). The EMU also gets the information regarding QoS from QoS control unit (QCU) and perform packet admission control accordingly. This framework extend the energy management feature by allowing QoS control unit to communicate with energy management unit so as to control the transmission power. This allows Layer 1 developers of the protocol stack to easily incorporate the design of transmission power control in a cognitive radio user.

V. Mishra et al., *QoS and Energy Management in Cognitive Radio Network*,
Signals and Communication Technology, DOI 10.1007/978-3-319-45860-1_6

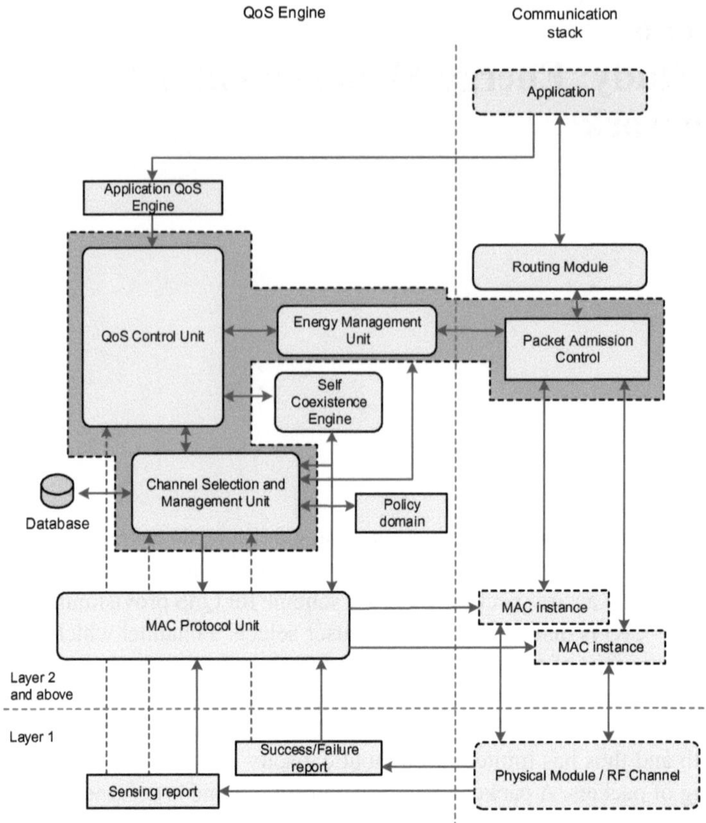

Fig. 6.1 Modules responsible for energy aware spectrum decision in the discussed QoS provisioning framework

6.1 System Model

The system comprises of a single cognitive radio network in *ad-hoc* mode, comprising N CR users each of which is equipped with a single half-duplex transceiver. The system operates in a heterogeneous environment comprising of C channels, each having bandwidth B_i, $\forall i \in \{1, ..., C\}$. The transmit power required on each channel is assumed to be proportional to the bandwidth of the channel and it is represented by P_i, $\forall i \in \{1, ..., C\}$. The spectral opportunity is homogeneous to all the CR users of the networks in terms of availability. Each of these channels is modeled as two state ON-OFF Markov chain as discussed in Sect. 3.4.1. Each CR node is also assumed to have a limited battery capacity. All CR users keep track of the available channels with an ideal channel sensing scheme. For the purpose of provisioning of QoS, the system classifies the user traffic into four different classes as discussed in Sect. 3.1.1.

6.2 Energy Aware Spectrum Allocation Scheme

The main assumption of the discussed scheme is that a smaller bandwidth channel will consume lower power for communication than its counterpart. The overall work of energy aware spectrum decision can be divided into two parts [2]:

1. Selecting a channel which utilizes low power.
2. Admission control on low priority packets.

The system first selects a channel satisfying the bare bandwidth need of the important applications (having higher priority). Once the bare bandwidth requirement is satisfied, the system performs admission control to the rest of the applications packets to utilize the rest of the bandwidth of the channel. The energy management unit (EMU) executes the threshold check for invoking energy aware spectrum allocation scheme in channel selection and management unit (CSMU). On receiving this command, the CSMU gets the information of bare bandwidth demand (discussed later) from QoS control unit (QCU) and then perform spectrum allocation accordingly which satisfies this bare bandwidth. Even with the allocation of the selected channel by higher priority packets, there is some bandwidth left on the channel through which lower priority packets can be transmitted. The packet admission control unit (PAC) allow scheduling and queuing of lower priority packets through remaining bandwidth of the channel.

6.2.1 Spectrum Decision

The energy management unit (EMU) keeps track of the power level and kicks in the energy aware spectrum decision at the channel selection and management unit (CSMU) when the power level drops below a certain threshold (user selectable). A guarantee on QoS in terms of bandwidth is provided to the top two applications: Voice and Video (from Table 3.1). Algorithm 6.1 provides the energy aware channel selection scheme which executes at the CSMU.

The channel indices used in Algorithm 6.1 is sorted in the order of bandwidth. Here *bare bandwidth* (ψ) is defined as the minimum bandwidth required to provide assured grade of QoS to the top two applications (voice and video packets). Therefore it can be assumed that the bare bandwidth is the bandwidth requirement of the top two applications and it can change depending upon the environment. A search is performed starting from lower bandwidth channels to find a channel which is available and has bandwidth greater than bare bandwidth demand i.e. $B_i \geq \psi$. The $band()$ function returns the bandwidth of the parameter passed. Here voice and video parameter represent the total number of voice and video packets respectively queued in the given time frame. To identify whether a channel is available or not, function $IsFree()$ is used. The function returns a value 1 when the channel ID passed is available otherwise it returns a value 0. If no channel satisfies the need (i.e. the

Algorithm 6.1: Algorithm for selecting a channel at a given time frame

```
Initialize : ψ = band(voice) + band(video);
Label:
for i = 1 to C do
    if band(channel(i)) ≥ ψ then
        if IsFree(channel(i)) then
            select_channel = i;
            break − for;
        end
    end
end
if select_channel = φ then
    ψ = band(voice);
    goto Label;
end
```

bandwidth ψ), then the bare bandwidth requirement is reduced to only voice packets. If even after this no channel is found to satisfy the requirement, then the currently selected channel is used to transmit low priority packets.

6.2.2 *Bare Bandwidth Calculation*

As discussed previously, bare bandwidth Ψ is calculated as sum of bandwidth requirement by *voice* and *video* packets. Within voice and video packets it is possible to have different access levels based on the waiting time of the packets in the queue. This can be followed by the fact that a video conference application has strict latency requirement for video packets than media applications like Youtube where packets can buffer if needed. Therefore to minimize the bandwidth use by different applications, the number of packet scheduling is reduced in such a way that packets with large latencies are scheduled first. Each application generating real time packet is assigned to have a maximum queue latency that it can suffer before performance degradation. This queue latency is given as δ_i for packets generated by video and voice applications, where i represents a packet ID. A_i and T represent the arrival time of packet i and current system time respectively. A packet to be considered for transmission in the current frame is selected using Algorithm 6.2 which executes at the QoS control unit (QCU) and is managed with the help of packet admission control unit (PAC).

The algorithm calculates the current latency of all the packets and selects only those which satisfy the given constraint in line 4. The expression $(T - A_{pkt})$ gives the current waiting time of the given packet (pkt). The expression $(T - A_{pkt} + FrameDuration)$ gives the maximum latency which the given packet will expect if it is deferred from transmission in the current time frame. Therefore the condition in

Algorithm 6.2: Algorithm for bare bandwidth calculation at a given time frame

```
1 Initialize : select_pkts = ϕ;
2 for i = 1 to Size(Q) do
3     pkt = Q_i ;
4     if (T − A_pkt + FrameDuration) > δ_pkt then
5         select_pkts = select_pkts ∪ {pkt};
6     end
7 end
8 ψ = band(select_pkts);
```

line 4 evaluates to true for packets which must be transmitted in the current time frame to meet the strict latency requirement and alleviate the system from performance degradation.

6.2.3 *Admission Control*

Once the channel is selected satisfying the bare bandwidth need, there is still some bandwidth left in the channel to allow the lower priority packets to pass through. An admission control technique is shown in Algorithm 6.3 to allow these packets to utilize the remaining bandwidth. This function is implemented in packet admission control unit of the framework. Even after transmission through the remaining bandwidth, some of the low priority packets are still blocked and hence need to be scheduled for transmission in next time frame. The discussed algorithm considers this scenario and attaches a counter to each blocked packet. The counter is initialized to $zero$ when the packet is blocked for the first time and is then incremented by a value of 1 whenever the packet is not sent in the next successive time frame. Whenever a packet selection is done, a packet having largest counter value is selected.

Algorithm 6.3: Algorithm for admission control to low priority packets in a given time frame

```
Initialize : λ = band(select_channel) − ψ;
num_best_pkt = (γ ∗ λ)/(best_pkt_size);
num_background_pkt = ((1 − γ) ∗ λ)/(background_pkt_size);

select_best_pkts = num_best_pkt from highest_counter(ζ);
select_background_pkts = num_background_pkt from highest_counter(ξ);
ζ = ζ − {select_best_pkts};
ξ = ξ − {select_background_pkts};
counter(ζ) = counter(ζ) + 1;
counter(ξ) = counter(ξ) + 1;
```

An assumption is made that the selected channel already satisfies the bare bandwidth requirement for all voice and video packets. Therefore low priority data like best effort data and background data (Table 3.1) can share the remaining bandwidth in a definite way. It can be assumed that the best effort data takes a share γ ($0 \leq \gamma \leq 1$) of the remaining bandwidth while background data takes $(1 - \gamma)$ share of the remaining bandwidth (λ). The algorithm first calculates the remaining bandwidth λ of the selected channel. The number of best effort packets and background packets are then calculated which can be allowed to pass through the channel (lines 2–3). The algorithm then selects these packets from the appropriate category having higher counter values compared to other blocked packets (lines 5–6). The set ζ represents the set of blocked best effort packets while set ξ represents the set of blocked background data packets (lines 7–8). The function $counter()$ is used to update the counter value of each packet in the set (lines 9–10). The set $select\_best\_pkts$ and $select\_background\_pkts$ consists of the packets allowed to pass through the current time frame.

6.3 Performance Analysis

The discussed spectrum decision scheme is compared against random channel selection scheme [3] due to lack of other comparable spectrum selection schemes with similar attributes. A random channel selection scheme is a channel selection technique in which a free channel is selected randomly and is used until that channel is sensed occupied by the primary user [3]. For a fair comparison, it is assumed that the random channel selection scheme selects a channel which satisfies the bare bandwidth demand of the node. The discussed scheme is evaluated using discrete event simulation in Matlab which simulates the network up to frame level. The media access scheme utilized for the performance evaluation is taken from Chap. 5. All the nodes in the network are assumed to be at one-hop distance to each other. The performance of the discussed scheme is evaluated in terms of energy consumption of the system and number of low priority blocked packets. The primary channels in the system are modeled as two state Markov chain as discussed in Sect. 3.4.1 with channel transition parameters α and β assumed to be uniformly distributed over the range [0, 1]. The bandwidth of the channel is modelled linear with the number of channels used, i.e. the minimum bandwidth of a channel is given a value of 0.5 Mbps and the maximum bandwidth of the channel is $0.5 * C$ Mbps, where C is the maximum number of channels used for the simulation. Therefore the bandwidth is distributed linearly from channel 1 to channel C as 0.5 to $0.5 * C$ Mbps. Similarly, the transmit power level on these channels is modeled to be linear with the number of channels used and its value varies from 0.1 to $0.1 * C$ watts. The number of voice communication packets, video communication packets, best-effort communication packets and background traffic communication packets used per second are 1, 2, 15 and 7 respectively giving the percentage ratio of 4, 10, 60 and 26 % for voice, video, best-effort and background traffic communication segments respectively. The size of voice, video, best-effort and background packets are 0.05, 0.5, 0.05 and 0.05 Mb respectively.

A relation between energy consumption and simulation time of the system is shown in Fig. 6.2. The number of primary channels used is 20 ($C = 20$). The discussed spectrum selection scheme (ESDF-CR) consumes 50 % less energy than random channel selection scheme. The result follows from the selection of channels which require low power for transmission and have satisfactory bandwidth. Thus the discussed energy aware spectrum decision scheme increases the operable life of the system.

A comparison between different low priority blocked data packets with respect to simulation time of the system is shown in Fig. 6.3. The number of primary channels used is 20 ($C = 20$). The total number of blocked best effort data packet and background data packet for ESDF-CR scheme increases faster than random channel selection scheme. This could be followed from the results in Fig. 6.2, where the discussed scheme selects low transmission power channels and thus utilizes lower bandwidth in comparison to that in the random channel selection scheme. The usage of low bandwidth channels results in blocking of low priority data packets.

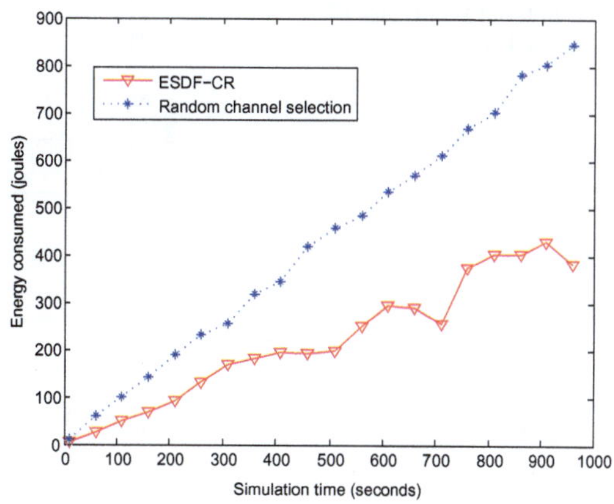

Fig. 6.2 Relation between energy consumption in the system with simulation time

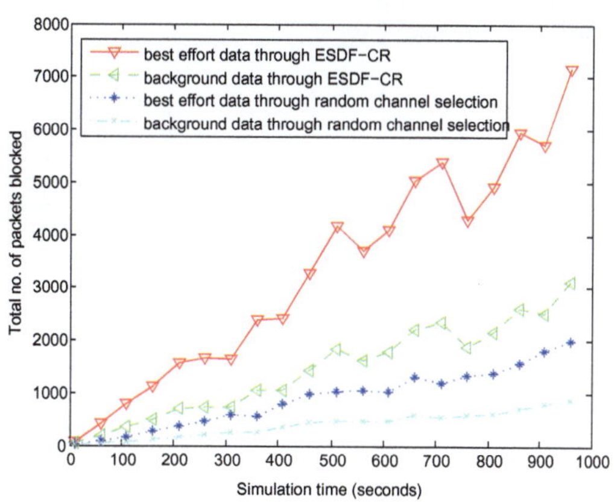

Fig. 6.3 Relation between number of packets blocked with simulation time

A relation between energy consumption of the system with the size of voice packet used by the application is shown in Fig. 6.4. The size of voice packet is varied from 0.05 to 0.5 Mb with the 20 primary channels ($C = 20$) and 100 s of simulation time. The size of other data traffic packets is kept unchanged. As the size of the voice packet increases the energy consumption also increases for ESDF-CR which can be followed from the requirement of higher bandwidth channels which consumes more energy. The effect of the variation of voice packet size can also be seen in random channel selection scheme where it tends to randomly select a channel which satisfies the bandwidth demand of varying size voice packet.

Figure 6.5 shows the relation between the number of blocked packets in the network with the size of voice packet used by the application for 20 primary channels ($C = 20$) and 100 s of simulation time. The size of voice packet is varied from 0.05 to 0.5 Mb. As the size of the voice packet increases the bare bandwidth demand also increases which results in selection of high bandwidth channels and higher energy consumption (Fig. 6.4). Since the bandwidth of channels used varies linearly with

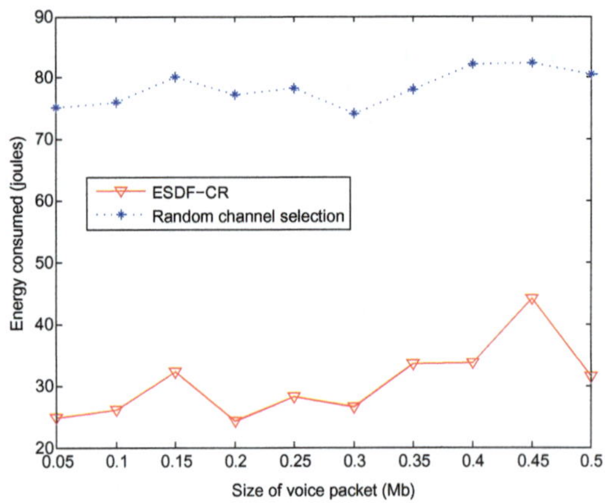

Fig. 6.4 Energy consumption versus size of voice packet

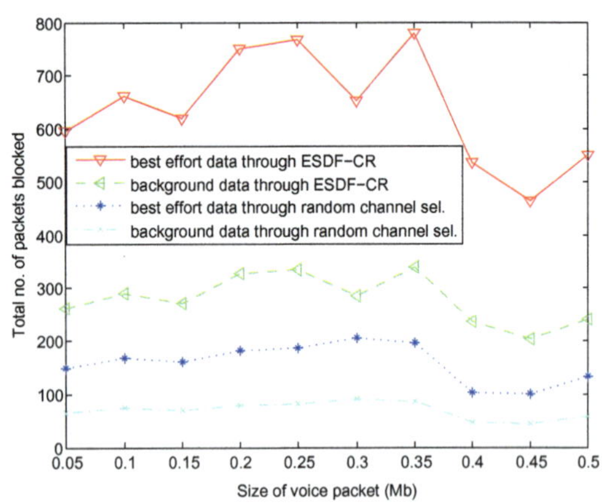

Fig. 6.5 Number of packets blocked versus size of voice packet

Fig. 6.6 Energy consumed versus total number of communication segments initiated

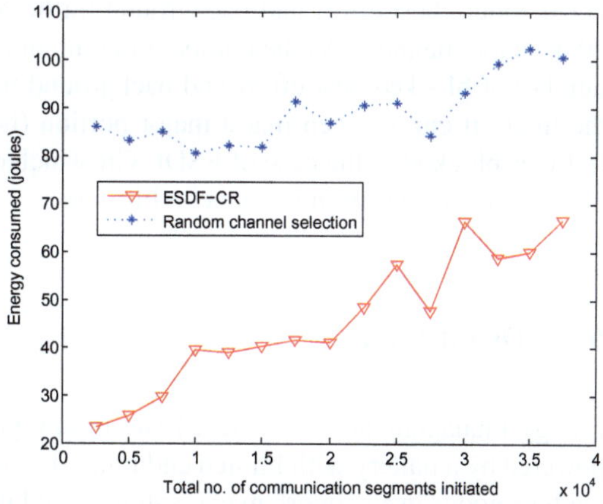

number of channels, and the number of voice segments are discrete in nature, a higher bandwidth channel will result in more unoccupied bandwidth after a certain threshold. Thus more best effort data and background data are allowed to pass resulting in decreased blocking rate.

A relation between total energy consumed in the system with the total number of data segments generated per 100 s is shown in Fig. 6.6. The size of voice, video, best effort and background traffic packets are all set to 0.05 Mb with 20 primary channels used in the system. The energy consumed by ESDF-CR increases with the number of communication segments used in the network. Also the discussed ESDF-CR starts to perform comparable to random channel selection scheme at higher data traffic. This can be followed from the selection of higher bandwidth channels to satisfy the bare bandwidth demand which increases with the traffic.

Figure 6.7 shows the relation between the number of blocked packets in the system with the total number of communication segments transmitted per 100 s. The size of

Fig. 6.7 Number of packets blocked versus total number of communication segments initiated

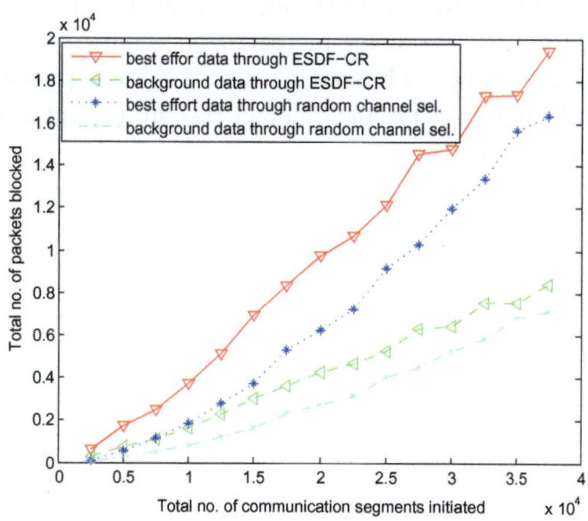

voice, video, best effort and background traffic packets are all set to 0.05 Mb with 20 primary channels. As the number of communication segments increases the total number of blocked best effort and background traffic packets also increases. From the figure it can be seen that a major portion (best effort and video traffic) of the traffic is blocked in the case of ESDF-CR which can be followed from the selection of low energy consuming channels (Fig. 6.6).

6.4 Discussion

Energy management is a critical issue in any portable system which is primarily powered by a battery with limited endurance between recharges. The failure of intermediate nodes in a network may result in route failure including critical loss of QoS. Energy management in a cognitive radio system can be done using three steps: *transmission power control, spectrum allocation,* and *media access scheme*. Transmission power control involves limiting the transmitter output power so as to decrease the power consumption along with the decrease in SNR/SINR. Energy consumption in a MAC scheme is managed with the help of etiquettes designed by the access scheme. Spectrum allocation based energy management involves selection of spectrum which consumes lower power for transmission by selecting channels with small bandwidth at lower frequency bands.

An energy aware spectrum decision scheme is discussed which allows cognitive radio network to enter into low power mode whenever the power of the system falls below a certain threshold. The scheme utilizes low power channels in low power mode to achieve the goal of increased network longevity. Techniques for low power channel selection along with admission control is discussed to realize the function of energy aware spectrum decision scheme. The traffic is divided into 4 classes to provide different access levels to different applications. The scheme selects a channel satisfying the assured grade of QoS of the top two priority traffic while admission control is applied on last two priority traffic. Performance analysis of the discussed scheme is done against random channel selection scheme and the result shows that the discussed scheme increases the longevity of the network while compensating for large number of blocked packets for lower priority applications. Once the spectrum selection is done, the channel IDs are passed to the media access scheme for their utilization.

References

1. Mishra, V., Tong Lau, C., Chan, S., & Mathew, J. (2013). Energy aware spectrum decision framework for cognitive radio network: A spectrum decision framework for cognitive radio network with energy awareness. *Journal of Low Power Electronics*, *9*(3), 313–321.
2. Mishra, V., Tong, L. C., Chan, S., & Kumar, A. (2012). Energy aware spectrum decision framework for cognitive radio networks. In, *International Symposium on Electronic System Design (ISED)* (pp. 309–313).
3. Bennai, M., Sydor, J., & Rahman, M. (2010). Automatic channel selection for cognitive radio systems. In, *IEEE 21st International Symposium on Personal Indoor and Mobile Radio Communications (PIMRC)* (pp. 1831–1835).

Chapter 7
Case Study: Self-coexistence in CRN Framework

The free use of wireless frequency spectrum by cognitive radio network is anticipated to increase its application in a multitude of wireless systems. The increase in use of such a system will result in a problem termed as *tragedy of commons* as discussed in Sect. 1.2.3. Since most of the cognitive radio systems will follow the same set of standards (and hence similar operating frequency spectrum), they will interfere if operated in vicinity of each other. The result will be a decrease in performance of the system throughput along with the degradation of QoS of individual cognitive user in the system. Mitigation of interference from similar homogeneous cognitive radio network is termed as self-coexistence [1].

As discussed in Sect. 1.8, the mitigation of mutual interference from other homogeneous cognitive radio network is a 3 step process involving: *resource relocation*, *transmission power control*, and *resource sharing*. Since the discussed QoS provisioning framework only considers layer 2 technology of the protocol stack, transmission power control for self-coexistence is beyond the scope of this book. Here, the self-coexistence scheme for cognitive radio network are discussed from the perspective of resource relocation and resource sharing.

7.1 Self-coexistence Procedure

Coexistence of multiple interfering homogeneous CR networks is realized with the help of Self-Coexistence Engine (SCE) of the discussed QoS provisioning framework. The work of self-coexistence engine can be broadly classified as: *detection* and *recovery* [2]. The SCE interacts with functional modules including success/failure report unit and MPU to detect interference among CR networks. While SCE interacts with CSMU, QCU and MPU to realize self-coexistence with QoS provisioning as shown highlighted in Fig. 7.1. The procedure for self-coexistence among interfering CR networks is discussed in following subsections.

V. Mishra et al., *QoS and Energy Management in Cognitive Radio Network*, Signals and Communication Technology, DOI 10.1007/978-3-319-45860-1_7

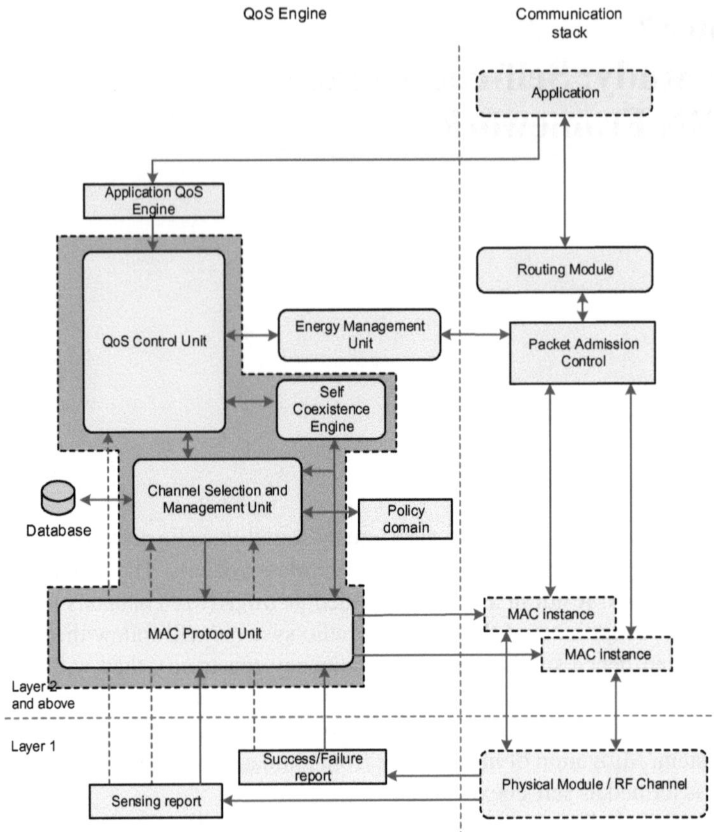

Fig. 7.1 Modules responsible for self-coexistence in cognitive radio networks

7.1.1 *Detection of Interfering Network*

When a user in a CR network receives a packet associated with another CR network (by checking network ID), it may conclude that it is in interference range of another CR network. A similar conclusion can be drawn if a user in a CR network receives corrupted packets for consecutive time periods. Once the interference is detected, the interfered user in the network requests for quiet period of multiple superframe duration on the channel in which it received the interference. During requested quiet period, users in the network attempt to listen to packets from interfering CR network/networks so as to deduce the network ID/IDs and their respective beacon start period. Once the beacon start period is known, an interfered CR network will perform inter-network communication to perform self-coexistence.

7.1.2 Recovery Process

As discussed in Sect. 1.8, an interfered user first attempts to relocate the operating spectrum band so as to mitigate the interference. If relocation of resource is not possible, then transmission power control (TPC) techniques are employed [1, 3]. If TPC techniques are unsuccessful, then the resource sharing based self-coexistence technique is applied. The resource (frequency spectrum) is shared among multiple interfering CR networks by having mutually exclusive access superframe of the channel. This is achieved by allocating each network with a non-overlapping superframes which are mutually exclusive. The recovery process via resource relocation and resource sharing are discussed in subsequent sections. Recovery process requires exchange of control messages among interfered and interfering networks along with exchange of control messages between interfered and interfering users. Therefore exchange of control packets can be classified as inter-network control messages and intra-network control messages.

7.2 Resource Relocation Based Self-coexistence

As discussed in Sect. 1.8, resource in the context of cognitive radio refers to a frequency band or a combination of time and frequency band. When multiple homogeneous cognitive radio networks operate in the vicinity, they may operate on same or similar set of resources leading to interference. Therefore the goal of self-coexistence from resource relocation perspective is to relocate the resources in such a way that the interfered cognitive radio networks operate on disjoint set of resources. Complexity wise, this method is more efficient and is less complex than other methods of self-coexistence.

For the purpose of QoS provisioning, the number of resources allocated to different networks may vary. This implies that in a resource constrained environment, a cognitive radio network with heavy bandwidth demand may consume all resources leading to resource starvation for cognitive radio network with small bandwidth demand [4]. Therefore, the system should be able to proportionally allocate resources to allow fairness while mitigating the problem of resource starvation. Resource allocation problems have been widely studied from different perspective including game theory, graph theory, heuristic algorithm, multi-objective optimization problem to name a few [5].

The problem of resource allocation can be directly mapped to the problem of multi-coloring of graph and hence in this scheme, multi-coloring of graph is considered to solve the problem of self-coexistence among multiple cognitive radio networks from the viewpoint of resource relocation [6]. Both, resource allocation problem and multi-coloring problem can be mapped to their corresponding optimization problems with a given objective and applied constraints. The *constraints* of the resource allocation problem directly maps to the *constraints* of the multi-coloring problem. Even with the availability of such mapping, no formal relation has been shown between the *objectives* of these optimization problems. A relationship between the objective of these optimization problems will allow deeper understanding about the design and analysis required to develop appropriate graph coloring heuristics. Such relationships are introduced in the subsequent sections. A multi-coloring heuristic for self-coexistence among multiple cognitive radio networks with QoS provisioning is also discussed. Design principles obtained from the development of relationship between the optimization problems of resource allocation and multi-coloring is utilized to develop the multi-coloring heuristic for self-coexistence.

7.2.1 System Model

The system comprises of multiple cognitive radio networks in ad-hoc mode. The total number of different cognitive radio network is M, each comprising of N_i cognitive users where $i \in \{1, \ldots, M\}$. The number of channels utilized in the system is C, each assumed to be of equal bandwidth B. The transmission range on each channel may vary, and hence a given pair of users may interfere on some channels while perfectly coexist on other set of channels. The set of interfered users are modeled through graph perspective as an undirected graph represented by $G = (V, E, L)$. V is the set of vertices representing the interfered users, and L ($L = \{L_v\}$) is the set of list of colors (L_v) representing channels available at a user $v \in V$. The set of edges representing interference between a user pair $u, v \in V$ on set of channels l is given as $E = \{(u, v,), \{l\}\}$, where $l \in \{1, \ldots, C\}$. For instance, $((x, y), \{1, 3\}) \in E$ implies that user x and y interfere on channels 3 and 4, while the group $((x, y), \phi) \in E$ implies that user x and y do not interfere on any channel and hence coexist with each other. An edge is present between two users if they interfere on at least one channel; therefore any edge marked with group $((x, y), \phi) \in E$ between vertices x and y can be assumed to be non-existent. Neighbor of a user v is the set of groups of user w and channels $\{l\} \subseteq \{1, \ldots, C\}$ over which user v and w interfere. This is given by $N(v) = \{(w, \{l|((v, w), \{l\}) \in E\}), v, w \in V, \{l\} \subseteq \{1, \ldots, C\}\}$. The expression implies that the neighbor of a user v is the set of groups of user w and channels $\{l\}$, where channels $\{l\}$ satisfy the condition $((v, w), \{l\}) \in E$. This work will use the terms channel and color interchangeably, similarly vertex and user are used interchangeably.

For instance, a graph theoretic perspective of a system of interfering cognitive radio based *ad-hoc* network is shown in Fig. 7.2. The system instance comprises of 6 interfering users from 3 different cognitive radio based ad-hoc networks operating over 4 channels $\{1, 2, 3, 4\}$. Users v_1 and v_4 are from network 1, users v_2 and v_6 are from network 2, while users v_3 and v_5 are from network 3. The interference between these users is shown via edges connecting these vertices and are marked with the colors over which the connecting users interfere. Therefore the system comprises of edges as $E = \{e_1, e_2, e_3, e_4, e_5, e_6, e_7\}$, while $e_1 = ((v_1, v_2), \{2\})$, $e_2 = ((v_1, v_3), \{3, 4\})$, $e_3 = ((v_2, v_3), \{1, \ldots, 4\})$, $e_4 = ((v_2, v_4), \{1\})$, $e_5 =$

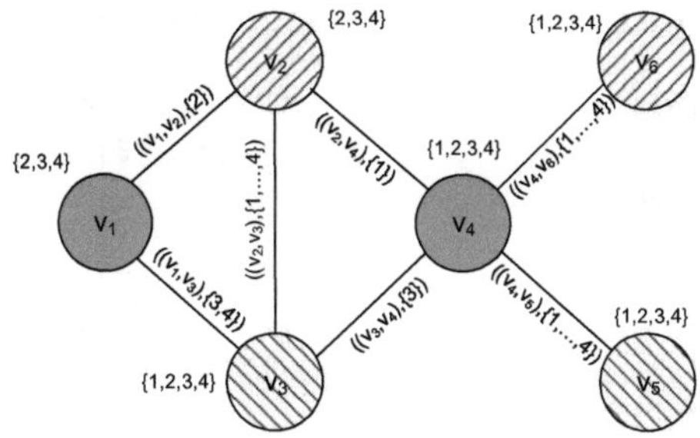

Fig. 7.2 Graph theoretic perspective of system of interfering cognitive radio based *ad-hoc* networks. The colors (channels) available at each user is shown along its corresponding vertex. The interference between users is represented via edges between vertices marked with the channels (colors) over which they interfere

$((v_3, v_4), \{3\}), e_6 = ((v_4, v_6), \{1, \ldots, 4\})$ and $e_7 = ((v_4, v_5), \{1, \ldots, 4\})$. The neighbor of user v_3 is given as $(v_1, \{3, 4\})$, $(v_2, \{1, \ldots, 4\})$, $(v_4, \{3\})$, (v_5, ϕ) and (v_6, ϕ).

7.2.2 Graph Coloring

In the previous section, mapping of the system of multiple cognitive radio network onto an undirected interference graph was discussed. In Fig. 7.2, an instance of a system is shown with the help of a graph, with available colors (channels) shown along the vertices. A multi-coloring of a graph G refers to the allocation of colors to the vertices following a certain constraint of rules. Therefore the multi-coloring of a graph is given be set $S = \{S_v\}$, where S_v is the list of colors allocated to vertex $v \in V$. By definition of traditional multi-coloring problem, S is a valid coloring of G iff $S_u \cap S_v \neq \{l\}, \forall((u, v), \{l\}) \in E$, where $u, v \in V$, $\{l\} \subseteq \{1, \ldots, C\}$, $S_u \in L_u$ and $S_v \in L_v$ [5]. Figure 7.3 shows two possible methods of multi-coloring of graph. The colors (channels) allocated are marked with '*' sign and are shown along the vertices.

Apart from the constraint discussed above for valid coloring, additional constraints can be imposed to achieve a certain goal. For a QoS provisioning network, additional parameter indicating the minimum number of channels required can be introduced. This is done so as to indicate the bandwidth demand of individual user in the system. A new class of problem is discussed, termed as edge restricted multi-coloring (ERMC) problem which includes an additional parameter of minimum number of channels required with the traditional multi-coloring problem. The additional parameter is given by $M = \{m_v\}$, where m_v is the minimum number of colors that need to be allocated to vertex $v \in V$. Therefore the ERMC problem can be formulated as:

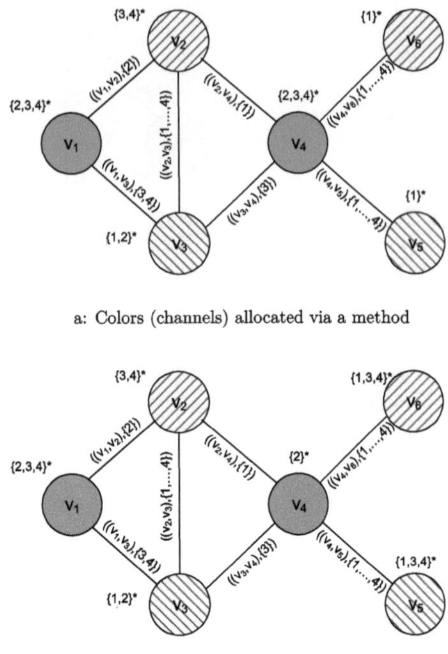

b: Colors (channels) allocated via an alternate method

Fig. 7.3 Colors allocated via different allocation strategies to the graph shown in Fig. 7.2. The colors allocated are shown via a star sign (*) along the vertices. Different allocation strategies result in different utilization of channels

Definition 7.2 Given a graph $G = (V, E, L, M)$, find a valid coloring of the graph with least number of different colors used, such that each vertex $v \in V$ in the graph is colored with at least m_v colors.

7.2.3 Optimization Problem Equivalent of ERMC

Let $N = |V|$ be the total number of vertices and C be the maximum number of different colors that are available in the system. Let the edges be represented by an $N \times N \times C$ matrix $E = \{e_{i,j,k}\}$, where $e_{i,j,k} = \{0, 1\}$. A value of $e_{i,j,k} = 1$ implies that an edge exists between vertices i and j and vice-versa. Therefore a value of $e_{i,j,k} = 0$ implies that vertices i and j may use the same color k. The list of available colors at each vertex is given by $N \times C$ matrix $L = \{l_{i,k}\}$. A value of $l_{i,k} = 1$ implies that color k is available at vertex i and vice-versa. The additional parameter representing minimum number of colors required at each vertex i is represented by a $N \times 1$ vector $M = \{m_i\}$. A value of $m_i = \delta$ implies that vertex i needs to be colored with at least δ number of colors. A valid coloring of the graph is represented by $N \times C$ matrix $S = s_{i,k}$. A value of $s_{i,k} = 1$ implies that color k is allocated to vertex i and vice-versa.

The equivalent optimization problem of the edge restricted multi-coloring (ERMC) problem can be formulated as:

$$
\begin{aligned}
\textit{minimize} \ & \sum_{k=1}^{C} v_k, \text{ where } v_k = \begin{cases} 1, \exists i[s_{i,k} = 1] \\ 0, \text{otherwise} \end{cases} \\
\text{subject to} \ & s_{i,k} \leq l_{i,k}, \\
& \sum_{k=1}^{C} s_{i,k} \geq m_i, \\
& s_{i,k}.s_{j,k}.e_{i,j,k} = 0, \\
& s_{i,k} = \{0, 1\},
\end{aligned}
\tag{7.1}
$$

for all $i, j = 1, \ldots, N$, $k = 1, \ldots, C$. Thus ERMC problem can be formulated as a non-linear integer optimization problem. This form of optimization problem is NP-hard in nature.

For channel allocation, the aim of a maximum throughput utility function is to allocate channels such that the throughput of the overall system is maximized. The optimization problem of this form of channel allocation is given as:

$$
\begin{aligned}
\textit{maximize} \ & \sum_{i=1}^{N} \sum_{k=1}^{C} s_{i,k} \\
\text{subject to} \ & s_{i,k} \leq l_{i,k}, \\
& \sum_{k=1}^{C} s_{i,k} \geq m_i, \\
& s_{i,k}.s_{j,k}.e_{i,j,k} = 0, \\
& s_{i,k} = \{0, 1\},
\end{aligned}
\tag{7.2}
$$

for all $i, j = 1, \ldots, N$, $k = 1, \ldots, C$.

Both optimization problems discussed in Eqs. 7.1 and 7.2 have equivalent constraints imposed except for different objectives. The equivalence of constraints allow solving the problem of resource allocation from graph theoretic perspective, except for the objective of the optimization problem.

7.2.4 Relationship Between Optimization Problems

Work done on resource allocation from the perspective of graph coloring exploit the fact that a solution to Eq. 7.1 will eventually satisfy all the constraints of Eq. 7.2. But it does not imply that the solution obtained from Eq. 7.1 is optimal for Eq. 7.2. This can be illustrated with the help of Fig. 7.3a, b where different coloring techniques have been applied with the additional parameter of minimum number of colors required by vertices as $m_1 = 3, m_2 = 2, m_3 = 2, m_4 = 1, m_5 = 1$ and $m_6 = 1$. Both resource allocation scheme results in same chromatic number[1] of 4, but have different throughput utility of 12 and 14 respectively. Therefore, a solution obtained from Eq. 7.1 implies minimum number of different colors used in the graph. Both Fig. 7.3a, b are in accordance with the Eq. 7.1 and hence are optimal solutions for it. But since they have different throughput utilities, only one of the figures will be optimal solution for Eq. 7.2, which in this case is Fig. 7.3b (since it has the highest throughput utility value).

Thus it can concluded that a solution of Eq. 7.1 may not be optimal for Eq. 7.2. It can be shown that a formal relation can be introduced between Eqs. 7.1 and 7.2 by adding the following constraint of maximum re-utilization of colors to Eq. 7.1.

$$maximize \sum_{i=1}^{N} s_{i,k}, \forall k \in K \tag{7.3}$$

Let the optimal solution of ERMC problem in Eq. 7.1 be K coloring (i.e. the chromatic number obtained is K). Now the optimization problem of maximum throughput utility function for channel allocation within K channels (colors) of the graph can be formulated as:

$$\begin{aligned} maximize \quad & \sum_{i=1}^{N}\sum_{k=1}^{K} s_{i,k} \\ \text{subject to} \quad & s_{i,k} \le l_{i,k}, \\ & \sum_{k=1}^{C} s_{i,k} \ge m_i, \\ & s_{i,k}.s_{j,k}.e_{i,j,k} = 0, \\ & s_{i,k} = \{0, 1\}, \end{aligned} \tag{7.4}$$

for all $i, j = 1, \ldots, N, k = 1, \ldots, C$.

Let the optimal solution of optimization problem in Eq. 7.1 along with the additional new constraint of Eq. 7.3 be S. Let an algorithm A be used to achieve this optimal solution. It can be proved by contradiction, that S is also an optimal solution to Eq. 7.2. Let S' be an optimal solution for Eq. 7.4 with K' coloring of its equivalent graph. Therefore

$$\sum_{i=1}^{N}\sum_{k=1}^{K} s_{i,k} < \sum_{i=1}^{N}\sum_{k=1}^{K'} s'_{i,k} \tag{7.5}$$

[1]Chromatic number refers to the minimum number of different colors required to color all the vertices in the graph following all the constraints imposed on the graph.

Two cases arise from this:

(i) When $K' < K$:This contradicts with the assumption that S is a solution for the optimization problem of Eq. 7.1. Since if there already exists a coloring which satisfies all constraints of Eq. 7.4 with $K' < K$, it would have been considered by Algorithm A.
(ii) When $K' = K$: From this we have $\sum_{i=1}^{N}\sum_{k=1}^{K} s_{i,k} < \sum_{i=1}^{N}\sum_{k=1}^{K} s'_{i,k}$ for a fixed value of K, $\implies \sum_{i=1}^{N} s_{i,k} < \sum_{i=1}^{N} s'_{i,k}$ which contradicts with the assumption in Eq. 7.3

Therefore solution S is optimal for Eq. 7.4. From this a more generalized optimization function of maximum throughput utility for channel allocation can be obtained. Suppose the coloring procedure is repeated for $K_1 + K_2 + \cdots K_n = C$ on disjoint sub-graphs $G_1, G_2, \ldots G_n$ of G for a finite n with K_1, K_2,...K_n all disjoint to each other. We obtain the following optimization formulation of maximum throughput utility function:

$$\begin{aligned} \textit{maximize} \quad & \sum_{i=1}^{N}\sum_{k=1}^{C} s_{i,k} \\ \text{subject to} \quad & s_{i,k} \le l_{i,k}, \\ & \sum_{k=1}^{C} s_{i,k} \ge m_i, \\ & s_{i,k}.s_{j,k}.e_{i,j,k} = 0, \\ & s_{i,k} = \{0, 1\}, \end{aligned} \tag{7.6}$$

Thus it can be concluded that an approximation algorithm which solves the ERMC problem with the intention of maximum re-utilization of all colors will subsequently approximately solve the optimization problem of maximum throughput utility function of channel allocation. This result is utilized in the next section for allocating channels based on their re-utilization.

7.2.5 *Self Coexistence Scheme with QoS Provisioning*

It can be safely assumed that nodes in the network support multiple priorities (categories) of data packet. Each node discovers its neighbor nodes and the channels on which it interferes with them. Here QoS is provisioned in terms of bandwidth and therefore a QoS metric is utilized for the purpose of calculation of bandwidth demand of each node. QoS metric is assumed to be the weighted average of last 10 frames in queue. Weighted average allows giving high weights to packets that were recently queued. It also allows prioritizing the bandwidth based on the packet category. The QoS metric (QM) of a node v in open spectrum system is calculated as:

$$QM_v = \frac{\sum_{j=0}^{9}\sum_{i=1}^{T} W_i.N_{ij}.(10-j)}{\left\{\sum_{i=1}^{T} W_i\right\}.\left\{\sum_{j=0}^{9}(10-j)\right\}} \tag{7.7}$$

Here N_{ij} represents the number of queued packet of type i in the last jth frame. T denotes the number of priorities (access categories) supported by the system. W_i are the weights assigned to packets of type i, where $i \in \{1,\ldots,T\}$.

For the purpose of fairness in channel allocation, a relative share of the demand for each node is calculated. A relative share indicates the bandwidth demand of a node in comparison to its neighbor interfering nodes. Since channel resources are re-utilized among all nodes of the interference graph, except among neighbor vertices; and neighbors are defined relative to the channel on which it causes interference, we calculate a resource share RS of a node v for channel i as:

$$RS_{v,i} = \frac{QM_v}{QM_v + \sum_{j\in N'(v)} QM_j}, i \in \{1, .., C\}, v \in V \tag{7.8}$$

where QM_v is the QoS metric of node v and QM_j is the QoS metric of neighbor nodes of v from unique network for channel i. Here $N'(v)$ is the set of neighbor node from unique network on channel i. Therefore *neighbor* node consists of only one node from a given network. For example in Fig. 7.2, node v_3 has neighbors as $(v_1, \{3, 4\})$, $(v_2, \{1, \ldots, 4\})$ and $(v_4, \{3\})$. For resource share calculation of node v_3 on channel 3, only nodes v_2 and v_4 are taken into consideration (since node v_4 and node v_1 are from the same network). This is done to avoid allocating more share to a network with higher number of interfered nodes. If two or more nodes from the same network, interfere on same channel, then the node which has the highest degree for all colors is considered for calculation. If two or more node satisfies this condition, then node having the least number of channels available is taken into account. If nodes have same number of channels available, then a node is randomly selected.

All nodes in the interference graph calculate a degree vector (DV), which is an element-wise summation of all channel availability lists (CALs) received from neighbor nodes including the CAL of node under consideration. Degree vector indicates the re-utilization of channels among the neighbors of a node. Each node $v \in V$ maintains a contention vector (CV) of size C, with each index consisting of a random value drawn from range

$$CV_i = \left[0, \frac{DV_i}{RS_i * \pi_i}\right], \forall i \in \{1, \ldots, C\} \tag{7.9}$$

where π_i is the steady state probability of channel $i \in \{1, \ldots, C\}$ to be free on node v (discussed in Sect. 3.4.1). The design of such contention value allows selection of channels with high availability and less usage (hence maximizes re-utilization) by nodes having high value of relative share; while randomness in selection of contention value allows mitigation of contention collision and resource deprivation. This contention vector is exchanged among all neighbors of the interference graph. The vertex then allocates all those channels for which the indices in contention vector has received least value.

A distributed algorithm for the implementation of the above discussed self coexistence scheme is shown in Algorithm 7.1. The given heuristic executes over each node $v \in V$ and approximately solve the ERMC problem with the probability of maximum re-utilization of colors with fairness. Calculation of QoS metric (line 1) allows QoS prioritization among nodes. The absoluteness in greed for prioritization is then removed by calculation of relative share *RS* (line 3). Degree vector (line 4) is calculated with the help of channel availability list (*CAL*). *CAL(j)* identifies the channel availability list of node j, whereas $CAL_i(j)$ implies the availability of channel i at node j. Contention vector is calculated and shared (line 5-7) so that vertices (nodes) win or lose colors (channels) on non-interfering basis. CV_i is the value of ith index of contention vector of node v, whereas $CV(w)$ is the contention vector of node w. Winning colors are then allocated to the vertex (line 8) and removed from channel availability list (line 9) of node v and its neighbors. The algorithm terminates on a node when all available channels on that node are allocated (lines 10-12). The algorithm has a polynomial bound convergence time for a node, given that the node or at least one of its neighbor node has a non-zero QoS metric.[2]

Algorithm 7.1: Distributed algorithm for self-coexistence executing over a node v

---

1 **Initialization** : For a node $v \in V$, Calculate and share QoS metric from Eq. 7.7;
2 **while** *true* **do**
3   Calculate *RS* from Eq. 7.8;
4   Calculate $DV_i = \sum_{j=1}^{N(v)} CAL_i(j), \forall i \in \{1, \ldots, C\}$
5     $CV_i = random[0, \frac{DV_i}{RS_i * \pi_i}], \forall i \in \{1, \ldots, C\}$;
6     $Transmit(CV(v))$;
7     $Receive(CV(w)), w \in \{N(v)\}$;
8     $colors = Select(min_i(CV(v), CV(w))), \forall i \in \{1, \ldots, C\}$;
9     $CAL(w) = CAL(w) \setminus \{colors\}, \forall w \in \{v, N(v)\}$;
10   **if** $CAL_v = \phi$ **then**
11   break-while;
12   **end**
13 **end**

---

[2]Proof is provided in Appendix D.

7.2.6 Performance Analysis

The performance of the discussed scheme is evaluated against traditional distributed list multi-coloring scheme [7] and utility graph coloring scheme [7] with frame level simulation in Matlab and ideal sensing scheme. For a fair comparison, the utility graph coloring scheme discussed in [7] is implemented in distributed way for ad-hoc network. The analysis is done in terms of convergence time, fairness in allocation and bandwidth deprivation of each node. Convergence time refers to the number of successive superframes required before all channels are allocated in the system. Bandwidth deprivation refers to average deprivation of bandwidth of each node and is calculated as the difference between bandwidth demand and bandwidth allocated to a node. Fairness refers to the unbiased allocation of resources (channels) to different nodes based on their proportional need (relative share) in the system. Jain's fairness index is utilized over bandwidth allocation ratio (ratio of bandwidth allocated to bandwidth demand of each node) [8]. Jain's fairness index for N nodes with x_i share of resource allocated to each node i is calculated as:

$$J(x_1, x_2, \ldots, x_N) = \frac{(\sum_{i=1}^{N} x_i)^2}{N * \sum_{i=1}^{N} x_i^2} \tag{7.10}$$

The simulation generates random interference complete graph for each run with uniform random multi-colored edges (interference). If an edge does not receive any coloring (i.e. no interference on any channel), then it is removed from the graph. To simplify simulation, all nodes in the graph are assumed to be from different networks. for all channels $i \in \{1, \ldots, C\}$ and for all vertices $v \in V$. Each channel utilized in the system is modeled as a two-state Markov chain as discussed in Sect. 3.4.1, and the parameters α and β are chosen as uniform random values in the range [0.1, 0.9] for all channels $i \in \{1, \ldots, C\}$ and for all vertices $v \in V$. Each node generates 4 types of Poisson modeled traffic: voice, video, best-effort and background; with percentage distribution of mean arrival as 4, 10, 60 and 26 % respectively. The size of voice, video, best-effort and background data packets used are 0.05, 0.5, 0.05 and 0.05 Mb respectively. Half of the nodes have mean data traffic as 4, 10, 60 and 26 packets per frame for voice, video, best-effort and background data packets; while the other half of the nodes have 16, 40, 240 and 104 packets per frame for voice, video, best-effort and background data packets. The bandwidth of each channel is assumed to be of 1 Mbps. The size of superframe used in the simulation is 100 ms and each data point in the graph is the average of 1000 runs.

Figure 7.4a, b shows convergence time (in terms of numbers of consecutive superframes) and fairness index (Jain's Index) with respect to total number of nodes used in the system. Total number of channels used is 100. The discussed coloring technique allows faster convergence with linear relation to that of number of nodes used in the system. This is attributed to the contention based design of resource allocation. On the other hand utility graph coloring scheme is derived from traditional graph coloring scheme and hence requires more convergence time than traditional coloring scheme

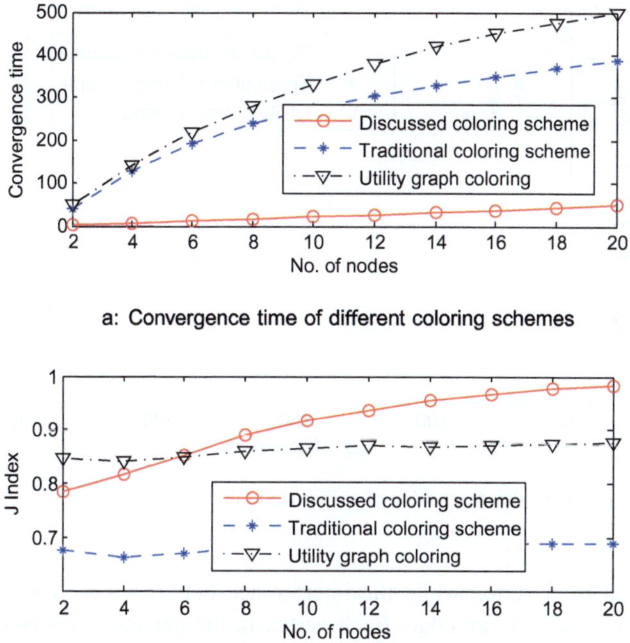

a: Convergence time of different coloring schemes

b: Fairness index of different schemes

Fig. 7.4 Convergence time and fairness index with respect to number of nodes

in *ad-hoc* environment. The discussed coloring scheme also provides fair proportional allocation (high value of J Index) to nodes in the system. This is attributed to the calculation and utilization of relative share (*RS*) in resource allocation (Eq. 7.8).

Figure 7.5 shows the average bandwidth deprived per node for a system with 8 nodes with respect to the number of channels used in the system. The number of channels used in the system is varied from 50 to 230. With the increase in number of channels, the decrease in deprived bandwidth can be attributed to the increase in number of channels allocated to a node in the system. The discussed coloring scheme allows proportional allocation and therefore has less deprived bandwidth in comparison to traditional allocation and utility graph coloring scheme.

7.2.7 *Infrastructure Mode Operation*

The resource relocation based self-coexistence technique discussed in the previous section utilized the system model of multiple networks in ad-hoc mode. The discussed scheme can be extended to support self-coexistence among multiple CR networks operating in *infrastructure* mode. To do so, each interfered cognitive access point

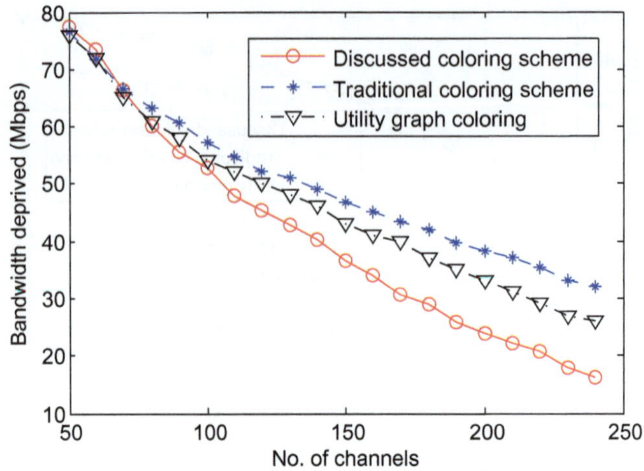

Fig. 7.5 Average bandwidth deprivation of each node in the system

is represented by a vertex while the interference between these cognitive access points is represented by an edge. Each vertex in the generated interference graph calculates its resource share by measuring and exchanging the QoS metric with its neighbors. Based on the resource share, the multi-coloring scheme discussed is utilized to allocate disjoint set of colors (channels) to each vertex in the interference graph.

7.3 Resource Sharing Based Self-coexistence

As discussed in Sect. 1.8, resource sharing based self-coexistence schemes can be utilized for the purpose of mitigation of interference between cognitive radio networks operating in the vicinity of each other. This method of self-coexistence is applied when resource relocation and transmission power control based self-coexistence are not successful to mitigate the interference. The complexity of self-coexistence based on resource sharing technique is higher than that in other techniques. Two methods of resource sharing have been discussed: *frequency* division multiple access and *time* division multiple access.

Since, *frequency* division multiple access based resource sharing is done at the Layer 1 of the protocol stack, it is beyond the scope of this book. In *time* division multiple access, a resource access (in this case duration of channel access) is divided into small time-durations. These time-durations are allocated in a disjointed way to all the interfering users such that each user has its own exclusive non-overlapping superframe. An access to non-overlapping superframe allows disjointed time-duration for channel access and hence mitigation of interference among multiple cognitive radio networks operating in the vicinity of each other [9]. The bandwidth allocated to each network is controlled by the size of superframe allocated to each CR network. In the subsequent sections, we discuss a QoS provisioning channel sharing based self-coexistence protocol for multi-channel CR based networks. The discussed scheme allows multiple homogeneous CR networks to coexist while minimizing interference and also guaranteeing a certain grade of QoS to the applications running on different nodes in networks.

7.3.1 System Model

The system model of multiple cognitive radio networks in ad-hoc network configuration is utilized. The open spectrum system comprises of M secondary *ad-hoc* networks, each comprises of N_i CR nodes, $\forall i \in M$. It is assumed that neighbor nodes follow common spectral opportunities. The system operates in an heterogeneous channel environment comprising of C channels, each of bandwidth B_i, $\forall i \in \{1, \ldots, C\}$. The system utilizes the MAC scheme discussed in Chap. 5 and the spectrum decision scheme discussed in Sect. 4.4 of Chap. 4. If the system operates over other set of MAC scheme and spectrum decision scheme, users in the system should have the capability to communicate the load distribution cross-layer in protocol stack. This allows allocation of appropriate bandwidth to different users in case a self-coexistence is required.

Different QoS metrics can be used depending upon the scheme used by the CR network. Here, a QoS metric is assumed to be the weighted average of packets sent/received in last 10 frames. The weighted average gives higher priority to transmission that happened recently. This is similar to the QoS metric defined in Sect. 7.2.5 and can be calculated as:

$$QM_v = \frac{\sum_{j=0}^{9} \sum_{i=1}^{T} W_i.N_{ij}.(10-j)}{\left\{\sum_{i=1}^{T} W_i\right\} . \left\{\sum_{j=0}^{9}(10-j)\right\}} \tag{7.11}$$

Here N_{ij} represents the number of queued packet of type i in the last j*th* frame. T denotes the number of priorities (access categories) supported by the system. W_i are the weights assigned to packets of type i, where $i \in \{1,\ldots,T\}$.

7.3.2 Design and Working

Since the discussed method of resource sharing is based on the sharing of time duration of channel access among different secondary users, the system is assumed to be divided into fixed length time frames. Any generic media access scheme with the capability of division of channel access into time-durations can be modified to

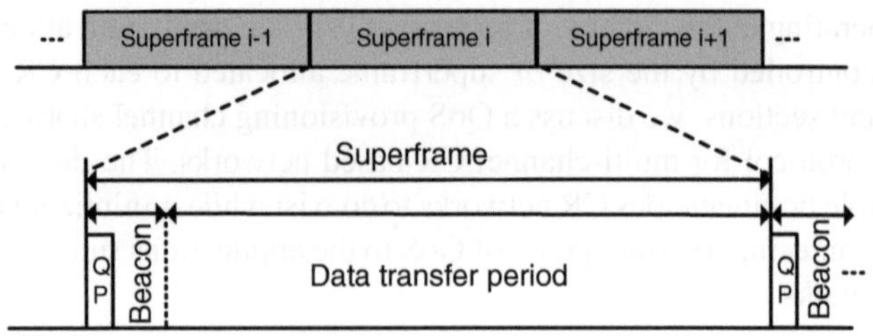

Fig. 7.6 Superframe structure of a medium access scheme

incorporate the discussed method of self-coexistence. Figure 7.6 shows a superframe of a generic medium access scheme comprising a beacon period and a data transfer period. During beacon period, all users in the network exchange beacon control messages and synchronize the whole network. During the data transfer period (DTP), users in the network perform the data transmission. Apart from these two durations, a cognitive radio network employs *quiet period* (QP) for the purpose of spectrum sensing. This period can be made as a part of beacon or it can be separately defined during data transfer period. Each network in the system is associated with a network ID (NID), which is embedded in all transmissions done by users in the network as a part of beacon.

7.3.2.1 Recovery Process

Once an interference is detected and the recovery process initiates resource sharing based self-coexistence, the interfering users starts communicating with each other via inter-network control messages. A resynchronization is required by one of the interfering CR network among two CR networks who are trying to coexist with each other, so as to mitigate any access collision. The CR network which resynchronizes is called as foreign network and a user in that network is called as foreign user. The network which does not perform any synchronization is called as home network and the user in it as home user.

Four inter-network control messages are employed for the purpose of self-coexistence:

- NWADV: Network advertisement message comprises of information about the network including number of channels, number of users, information about operating channels, coexisting network information, beacon start period, number of users on the current channel, quiet period distribution etc. The information about the QoS metric is also encoded in the packet so as to enable interfering CR networks to calculate the size of exclusive superframe duration for the purpose of QoS provisioning.

- NWACK: Network acknowledgement message serves as an acknowledgement to the NWADV messages. This message ensures connectivity between coexisting CR networks.
- SYN: Synchronization message is sent from a home user to a foreign user to indicate that the foreign network has to resynchronize with the home network.
- SYNACK: Synchronization acknowledgement is sent by a foreign user to the home network to indicate that the foreign network has resynchronized with the home network and is capable of coexisting with the home network.

Figure 7.7 shows a self-coexistence session via the help of inter-network control messages. For resource sharing based self-coexistence, when a user detects interference, it waits for a random duration of time and sends NWADV message to the interfering user/network. This random duration allows mitigation of collision of NWADV messages which may be initiated by the other CR network. Once the NWADV message is transmitted, the user waits for his acknowledgement (NWACK). Based on the exchange of NWADV, a decision is made about which networks will act as the home network and the foreign network. Once the decision is made, each interfered network calculates its superframe size, and this superframe size is passed by the home user to the foreign user along with a request to resynchronize its network. Once the foreign network is synchronized with the home network, the foreign user sends an acknowledgement (SYNACK). The home network verifies the synchronization through the help of beacon start period and superframe boundaries.

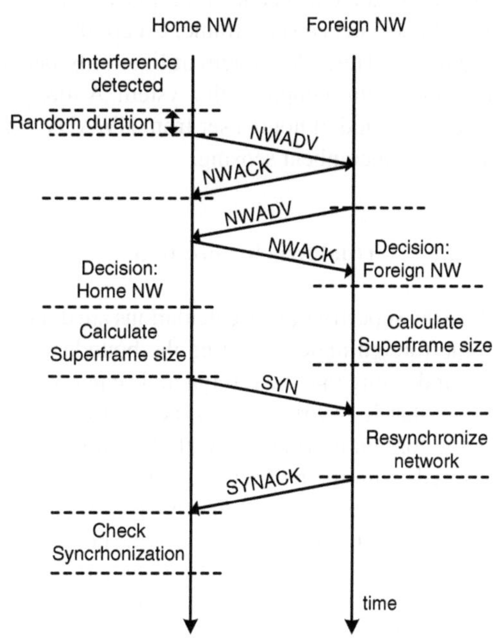

Fig. 7.7 Exchange of inter-network control packets for self-coexistence

7.3.2.2 Home/Foreign Network Decision

For a coexistence procedure to operate, the system has to decide which CR networks will act as the home network and which as the foreign network. During the exchange of NWADV messages, if the message indicates that it is already coexisting with some other CR network, then the network with whom a new network interferes is selected as the home network. If none of the network already coexists with any other networks, then the network which has largest number of nodes is chosen as the home network. This is done to decrease the convergence time required by foreign network for resynchronization.

7.3.2.3 Superframe Size Calculation

An interfered user in the network calculates the size of the superframe by utilizing the *QoS_Metric* field of NWADV message. For two CR networks which are trying to coexist with each other, a home user calculates the size of superframe as:

$$Size = \left(\frac{QM_H}{QM_H + QM_F}\right) * (SF - QP) + QP \tag{7.12}$$

QM_H is the QoS metric of the home user while QM_F is the QoS metric of the foreign user. QP is the size of the quiet period and SF is the original superframe size before the coexistence took place. Similarly, a foreign user also calculates the size of coexistence superframe, except the size of quiet period is removed from the calculation. Here, the system utilizes only one quiet period for whole time frame to increase the throughput of the system. A foreign user learns about the distribution of the quiet period of home user from NWADV message and utilizes it as its own quiet period for incumbent sensing.

7.3.2.4 Virtual Synchronization

Once the superframe size calculations are done, the home node sends a SYN message to the foreign node indicating the boundary values of its own superframe size and also indicating that the foreign node can now resynchronize its own network in such a way that it is virtually in sync with the home network. If the calculations done by the foreign node agrees with the values sent by the home node, the foreign node proceeds with the resynchronization of the beacon start period using intra-network control packets. Once the foreign network is resynchronized, the foreign node sends back a SYNACK message to the home node indicating that it is virtually synchronized with the home network. To check the virtual synchronization, the home node checks for foreign network beacon to start at the end of the home network superframe. Similar check is done by the foreign node to ensure that there are no collisions in frame durations. The exchange of inter-network control packets between two CR

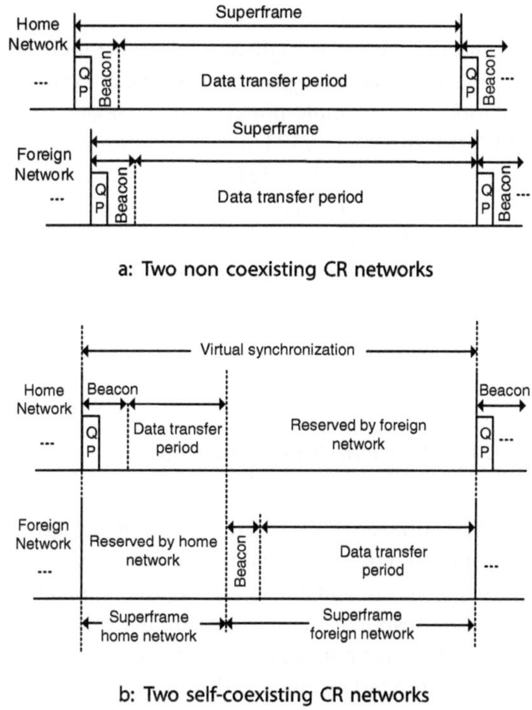

Fig. 7.8 Coexistence procedure

network is shown in Fig. 7.7. A virtual synchronization process is shown in Fig. 7.8 with two homogeneous CR networks before coexistence procedure (Fig. 7.8a) and after coexistence procedure (Fig. 7.8b).

7.3.3 Multiple CR Network Coexistence

Whenever nodes of multiple CR networks want to coexist with each other, they co-exist sequentially one by one. Waiting for random duration of time before transmitting NWADV message helps in avoiding collisions between multiple NWADVs. If a node overhears a co-existence initiation session going on between two other CR networks, it waits for SYNACK message and then initiates its own co-existence session. Whenever a new node wants to join, all previously joined nodes have to perform the calculation of superframe size. An example calculation of superframe for *jth* foreign network can be done as following:

$$Size = \left(\frac{QM_{Fj}}{QM_H + \sum_{i=1}^{M} QM_{Fi}} \right) * (SF - QP) \quad (7.13)$$

Here M is the number of CR networks which want to coexist with each other. All new joining networks will resynchronize their beacons to meet the boundaries of previously joined foreign networks. The coexistence is done between only interfering networks. For example if a system has nodes from three CR networks $NW1$, $NW2$ and $NW3$ in which $NW1$ interferes with $NW2$ while $NW2$ interferes with $NW3$, then co-existence is done between $NW1 - NW2$ and $NW2 - NW3$. The node of CR network $NW2$ propagates the information that it already coexist with some other CR network and would ask the joining network to act as foreign network. The joining node would join considering the boundary value of $NW2$. Here $NW2$ acts as a home network for communication of inter-network control messages with the new joining network.

7.3.4 Performance Analysis

The performance of the discussed resource sharing based self-coexistence technique is evaluated in terms of real time connection failure with respect to total number of communication segments used and total number of channels used in the system. The discussed resource sharing scheme is evaluated using discrete event simulation in Matlab which simulates the network up to frame level. All C channels are modeled as two-state Markov chain as discussed in Sect. 3.4.1, and the parameters α and β for channel availability are assumed to be uniformly distributed over the range [0, 1]. The bandwidth of each channel is modeled to be linear with the number of channels used in the system, i.e. the minimum bandwidth of a channel is given a value 1 Mbps and the maximum bandwidth of the channel is C Mbps. Therefore the bandwidth is distributed linearly from channel 1 to channel C as 1 to C Mbps. The system utilizes the QoS provisioning medium access scheme discussed in Chap. 5 with four access categories. Therefore the system has 4 types of packets: voice, video, best effort and background data. To simplify the analysis, it can be assume that two mutually interfering homogeneous CR networks $NW1$ and $NW2$ are trying to coexist with each other on all operating channels via resource sharing approach. All channels have a pair of communicating nodes from each network. The traffic from these nodes consists of a mixture of voice, video, best-effort and background data with percentage distribution as 4, 10, 60 and 26 % respectively [10]. The size of voice, video, best-effort and background packets are 0.05, 0.2, 0.05 and 0.05 Mb respectively. For performance analysis in terms of QoS, it is assumed that network $NW2$ has 3 times more traffic intensity than that of network $NW1$. The total simulation time is 100 s, whereas each point in the graph is the average run of 100 tests. The terms- connections, packets and segments are used interchangeably.

Figure 7.9 shows the number of voice connections failed with respect to the total number of voice segments initiated. The total number of channels used in the system is 10 whereas the total number of voice segments initiated is varied from 25 to 500 per second. The number of voice connections failure increases with the increase in total number of voice segments due to the bandwidth demand exceeding the

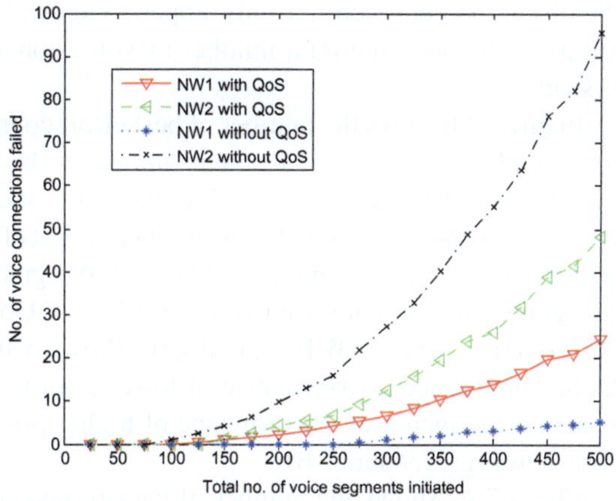

Fig. 7.9 Number of voice connections failed with increase in total number of communication segments

available bandwidth. Networks which coexist without QoS provisioning have very small number of voice connection failures and large number of voice connection failures for $NW1$ and $NW2$ respectively. Therefore the system appears to be biased towards the network having small traffic intensity ($NW1$). The discussed resource sharing scheme with QoS provisioning distributes the bandwidth according to the traffic intensity, therefore the disparity between the number of voice connection failed for both coexisting network is reduced.

Figure 7.10 shows the total number of voice connection failed including $NW1$ and $NW2$ with respect to total number of voice segments initiated. The total number of channels used in the system is 10 whereas the total number of voice segments initiated is varied from 25 to 500 per second. As the total number of voice segments increases, the discussed approach reduces the number of voice connection failures for the combined system of multiple networks. As evident from Figs. 7.9 and 7.10,

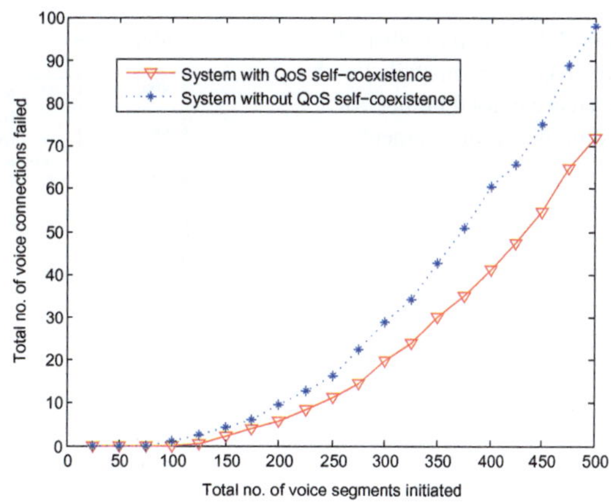

Fig. 7.10 Total number of voice connections failed with increase in total number of communication segments

the discussed approach not only imparts bandwidth distribution according to load but also decreases the total number of voice connections failure for the combined system.

Figure 7.11 shows the number of best effort connections failed with respect to the total number of best effort segments initiated. The total number of channels used in the system is 10 whereas the total number of best effort segments initiated is varied from 375 to 7500 per second. The number of best effort connections failure increases with the increase in total number of best effort segments due to the bandwidth demand exceeding the available bandwidth. $NW2$ with QoS and without QoS shows equal performance, while $NW1$ without QoS shows better performance than $NW1$ with QoS. This reduced performance in lower priority packets of QoS based networks is due to the increased performance of higher priority packets in QoS networks as shown in Figs. 7.9 and 7.10.

Figure 7.12 shows the number of background connections failed with respect to the total number of background segments initiated. The total number of channels

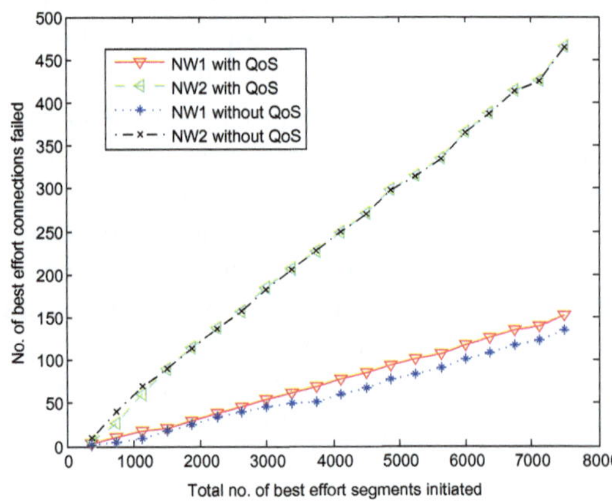

Fig. 7.11 Number of best effort connections failed with increase in total number of communication segments

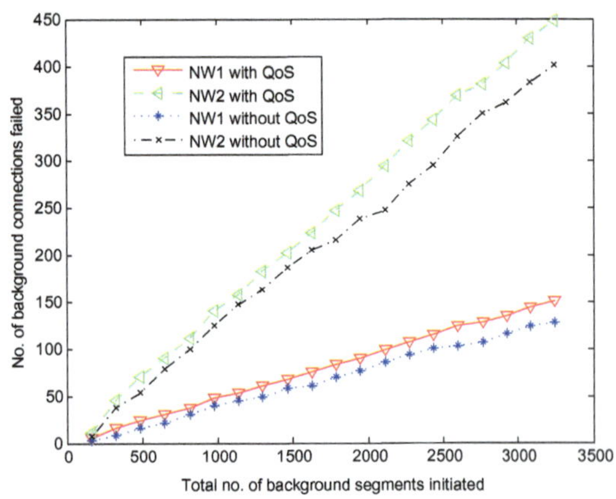

Fig. 7.12 Total number of voice connections failed with increase in total number of communication segments

used in the system is 10 whereas the total number of background segments initiated is varied from 163 to 3250 per second. The number of background connections failure increases with the increase in total number of background segments due to the bandwidth demand exceeding the available bandwidth. Networks $NW1$ and $NW2$ without QoS performs better than networks $NW1$ and $NW2$ with QoS. This reduced performance in lower priority packets of QoS based networks is due to the increased performance of higher priority packets in QoS networks as shown in Figs. 7.9 and 7.10.

In Fig. 7.13, an analysis of the total number of voice connection failed with the number of channels used in the system is shown. The traffic consists of 15, 37, 225 and 95 communication segments of voice, video, best-effort and background data which gives the percentage distribution as 4, 10, 60 and 26 % for voice, video, best-effort and background data respectively. The number of channels is varied from 2 to 22. From the graph, it can be observed that the number of channels have no effect on the number of voice connection failures in the networks. Also the discussed approach continues to show a decrease in disparity between the connection failures of home and foreign network.

In Fig. 7.14, an analysis of the total number of voice connections failed with the total number of channel used in the system is done. The traffic consists of 15, 37, 225 and 95 communication segments of voice, video, best-effort and background data which gives the percentage distribution as 4, 10, 60 and 26 % for voice, video, best-effort and background data respectively. The number of channels is varied from 2 to 22. From the graph, it can be observed that the discussed approach reduces the combined number of voice connection failure in the system.

Fig. 7.13 Number of voice connections failed with different number of channels used

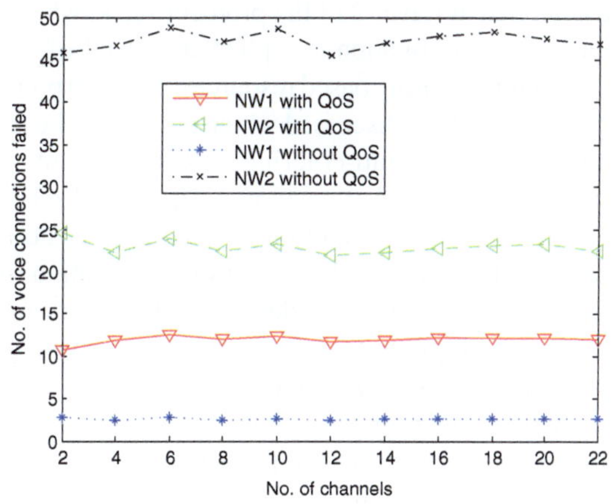

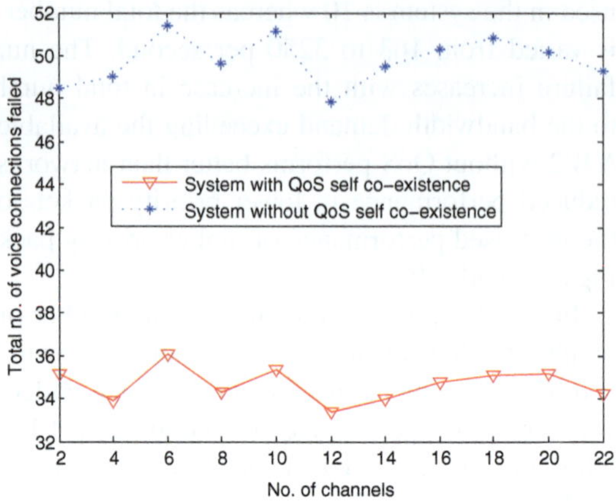

Fig. 7.14 Total number of voice connections failed with different number of channels used

7.4 Discussion

The open spectrum system promises free usage of spectrum on an opportunistic basis. But this free usage will eventually suffer from a problem termed as *tragedy of commons*, which will increase the interference among cognitive radio based wireless system and thus restricting their growth. To mitigate such scenario, self-coexistence techniques have been introduced which mitigates interference among similar homogeneous cognitive radio networks. Three different forms of self-coexistence techniques exists: *resource relocation, transmission power control* and *resource sharing*. Resource relocation and resource sharing based interference mitigation techniques are applied at Layer 2 of the protocol stack, while transmission power control based interference mitigation is applied at Layer 1 of the protocol stack.

A multi-coloring based resource relocation policy for self-coexistence in cognitive radio networks is discussed which allows for the formation of a relationship between edge restricted multi-coloring (ERMC) scheme and maximum throughput utility function of channel allocation. A distributed scheme for solving ERMC problem is discussed with the intention of maximum re-utilization. The multi-coloring based resource relocation policy allows fair allocation of channels while serving the QoS demands of individual nodes. It has been shown that the discussed scheme allows fairness with proportional allocation and converges much faster than utility graph coloring and traditional coloring scheme.

A frame sharing based self-coexistence procedure for cognitive radio networks is also discussed from the perspective of resource sharing. The discussed scheme allows generic medium access scheme to incorporate the feature of self-coexistence using little modification. The scheme allows two or more homogeneous CR networks to virtually synchronize and share media with each other on a non-interfering basis. The sharing is done using modified superframe size which depends upon the QoS requirement of the interfered node of a CR network. The frame sharing based approach for resource sharing allows fair calculation of frame size while serving the

QoS demands of individual users. It has been shown that the discussed scheme not only impart bandwidth distribution according to load but also decreases real time connections failure of combined system.

References

1. Lee, D. H., Jeon, W. S., Jeong, D. G., & Ko, G. (2010). Self-coexistence techniques for cognitive radio lans/pans. In *IEEE 21st International Symposium on Personal Indoor and Mobile Radio Communications (PIMRC)*, (pp. 1516–1520).
2. Vishram, M., Tong, L. C., & Syin, C. (2015). A channel allocation based self-coexistence scheme for homogeneous ad-hoc networks. *IEEE Wireless Communications Letters*, *4*, 545–548.
3. Hoven, N., & Sahai, A. (2005). Power scaling for cognitive radio. In *International Conference on Wireless Networks, Communications and Mobile Computing*, *1*, (pp. 250–255).
4. Pattara-Aukom, W., Banerjee, S., & Krishnamurthy, P. (2002). Starvation prevention and quality of service in wireless lans. In *The 5th International Symposium on Wireless Personal Multimedia Communications*, *3*, (pp.1078–1082).
5. Han, R., & Liu, K. (2008). *Resource allocation for wireless networks: basics, techniques, and applications*. Cambridge: Cambridge University Press.
6. Vishram, M., Tong, L. C., & Syin, C. (2014). List multi-coloring based fair channel allocation policy for self coexistence in cognitive radio networks with qos provisioning. In *Region 10 Symposium, 2014 IEEE*, (pp.99–104).
7. Sengupta, S., Brahma, S., Chatterjee, M., & Shankar, N. S. (2013). Self-coexistence among interference-aware IEEE 802.22 networks with enhanced air-interface. *Pervasive and Mobile Computing*, *9*(4), 454–471.
8. Jain, R., Durresi, A., & Babic, G. (1999). Throughput fairness index: An explanation, Technical Report, Department of CIS, The Ohio State University.
9. Vishram, M., Tong, L. C., & Syin, C. (2013). Qos provisioning self co-existence protocol for cognitive radio ad-hoc networks. In *19th IEEE International Conference on Networks (ICON)*, (pp.1–6).
10. Mishra, V., Tong, L. C., & Syin, C. (2012). Qos based spectrum decision framework for cognitive radio networks. In *18th IEEE International Conference on Networks (ICON)*, (pp.18–23).

Appendix A
Proof of Lemma 4.2

Algorithm 4.1 based on the greedy approach of selecting the longest consecutive free time-slots provides an optimal solution to the minimal channel switching problem. We provide a proof by contradiction by considering a vector **V**, which is obtained using the discussed greedy approach mentioned in the Algorithm 4.1. Let there exist a vector **U** which is not obtained by the greedy approach and has

$$\sum_{k=0}^{\eta-1} Switching(\mathbf{U_k}) \leq \sum_{k=0}^{\eta-1} Switching(\mathbf{V_k}) \tag{A.1}$$

where $\eta = (24 * 60/x) - 1$ and x is the sampling interval in minutes.

Also, $size(\mathbf{U}) = size(\mathbf{V}) =$ total time-units

$$= (24 * 60/x) - 1 = constant \tag{A.2}$$

Let m and n represents the total number of channel switching required when the system utilizes vector **U** and **V** for channel selection respectively. Let α_i and γ_j represent the vector indices blocks which have consecutively same channels in vector **U** and **V** respectively; the values i and j identify a particular block in vector **U** and **V** respectively. The values of i and j range from 1 to $m+1$ and $n+1$ respectively. We assume that α_i = 1 and γ_j = 1 for all possible values of i and j. We also define time blocks t_i and θ_j as the total time-units to which the system remains in blocks α_i and γ_j respectively. For example, if channel selection of a vector **U** is given by [1 1 3 3 2 2 2 3 3], then the total number of switching $m = 3$; total time-units in blocks $t_1 = 2$, $t_2 = 2$, $t_3 = 3$ and $t_4 = 2$. Now from Eq. A.2

$$size(\mathbf{U}) = size(\mathbf{V}) \implies \sum_{i=1}^{m+1} \alpha_i t_i = \sum_{j=1}^{n+1} \gamma_j \theta_j \tag{A.3}$$

From Eqs. A.1 and A.3 $\implies \quad m < n$

V. Mishra et al., *QoS and Energy Management in Cognitive Radio Network*, Signals and Communication Technology, DOI 10.1007/978-3-319-45860-1

Now, if $m < n$ and $\alpha_i = \gamma_j = 1, \forall i \in \{1, \ldots, m+1\}, j \in \{1, \ldots, n+1\}$, the given equality in Eq. A.3 $\implies \exists i \exists j : t_i > \theta_j$.

This contradicts with the assumption that the vector $\mathbf{V}$ is obtained using greedy approach. If there exists any time block t_i greater than θ_j, then that would have been taken by our greedy approach.

Appendix B
Proof of Lemma 4.3

To prove that High Throughput Minimal Switching (HTMS) problem is in NP, we show that a solution to this problem can be verified in polynomial time. Let an instance of solution be vector V. To check whether it satisfies the constraints (total number of switching $< \delta$ and bandwidth $> \omega$) or not, we need to check:

(i) $\sum_{k=0}^{\eta-1} Switching(V_k) < \delta$
(ii) $\sum_{k=0}^{\eta} BW(V_k) > \omega$, where $BW(i)$ returns the bandwidth of channel i
(iii) $\mathbf{M}_{\mathbf{V_k},\mathbf{k}} = 1, \forall k \in \eta'$, where $\mathbf{M}$ is channel availability matrix

Verifying

(i) Takes a running time of $O(\eta)$
(ii) Takes a running time of $O(\eta + 1)$
(iii) Takes a running time of $O(\eta + 1)$

Thus an instance of the solution can be verified in polynomial time.

V. Mishra et al., *QoS and Energy Management in Cognitive Radio Network*,
Signals and Communication Technology, DOI 10.1007/978-3-319-45860-1

Appendix B
Proof of Lemma 4.3

Appendix C
Proof of Lemma 4.4

To prove that High Throughput Minimal Switching (HTMS) problem is NP-hard, we will reduce a known NP-hard problem into HTMS problem. We consider Linearly Constrained Binary Quadratic Programming (LCBQP) which is a known NP-hard problem [1, 2]. Let the LCBQP be of the form:

$$f(x) = x^t Qx = \sum_{i=1}^{n}\sum_{j=1}^{n} q_{i,j}x_i x_j \qquad x_i \in \{0, 1\}\forall i \tag{C.1}$$

subject to $a'x < b$.

We modify this LCBQP problem by considering vectors consisting of binary values instead of variables x_i, x_j etc., which makes it easier to reduce it into our problem. Let the solution to HTMS problem be represented by a vector **VS**. We convert this solution vector into solution matrix **MS**, which consists of rows and columns representing the channels ids and time-units respectively. A value of one inside the matrix means that the corresponding row is selected for that column i.e. a particular channel is selected for that particular time-unit.

i.e.

$$\mathbf{MS}_{i,j} = \begin{cases} 1, & \text{if } \mathbf{V}_j = i \\ 0, & \text{otherwise} \end{cases} \forall i \in C', \forall j \in \eta' \tag{C.2}$$

where i, j represent rows and columns for matrix **MS**; similarly, for vector **V**, j represents its index. Let $\mathbf{X}_j$ be the vector representing the jth column of the solution matrix **MS**. Now we formulate the objective function and various constraints. We define the objective function to be minimum switching, satisfying a given bound on channel bandwidth.

The minimum switching is given by maximizing the following quadratic function.

$$\begin{aligned} Switching =& \mathbf{X}_0^T.\mathbf{X}_1 + \mathbf{X}_1^T.\mathbf{X}_2 + \mathbf{X}_2^T.\mathbf{X}_3 + \mathbf{X}_3^T.\mathbf{X}_4 + \\ & \mathbf{X}_4^T.\mathbf{X}_5 + \ldots \mathbf{X}_\eta - 1^T.\mathbf{X}_\eta \end{aligned} \tag{C.3}$$

V. Mishra et al., *QoS and Energy Management in Cognitive Radio Network*, Signals and Communication Technology, DOI 10.1007/978-3-319-45860-1

We assume each channel has bandwidth B_c, where c is the channel index. We assume a vector **BV** of size C, consisting of bandwidths of all channel and is given as:

$\mathbf{BV} = [B_1 B_2 B_3 B_4 B_5 B_6 B_7 \ldots B_C]$

Therefore, overall bandwidth for 24 h is given as,

$$\begin{aligned} BW = &\mathbf{BV}.\mathbf{X}_0 + \mathbf{BV}.\mathbf{X}_1 + \mathbf{BV}.\mathbf{X}_2 + \mathbf{BV}.\mathbf{X}_3 + \mathbf{BV}.\mathbf{X}_4 + \\ &\mathbf{BV}.\mathbf{X}_5 + \ldots \mathbf{BV}.\mathbf{X}_\eta \end{aligned} \tag{C.4}$$

We define bandwidth BW to be greater than some constant ω. Now we describe the validity constraints of the system. For a solution vector **VS** to be valid, it needs to satisfy three constraints:

(i) The vectors $\mathbf{X}_0, \mathbf{X}_1, \mathbf{X}_2, \mathbf{X}_3, \mathbf{X}_4, \mathbf{X}_5 \ldots$ etc. are valid only if these satisfy the availability of channels in channel availability matrix **M**. We define $\mathbf{M}_j$ as the vector of column j of matrix M. Therefore from this constraint we have

$$\begin{aligned} &\mathbf{M}_0^T.\mathbf{X}_0 + \mathbf{M}_1^T.\mathbf{X}_1 + \mathbf{M}_2^T.\mathbf{X}_2 + \mathbf{M}_3^T.\mathbf{X}_3 + \mathbf{M}_4^T.\mathbf{X}_4 + \\ &\ldots \mathbf{M}_\eta^T.\mathbf{X}_\eta = \eta + 1 \end{aligned} \tag{C.5}$$

(ii) There could be only one index having a value of 1 in each solution vector X. Therefore from this constraint we have,

$$\mathbf{1}_\eta^T.\mathbf{X}_j = 1, \forall j \in \eta' \tag{C.6}$$

where $\mathbf{1}_\eta$ is a row vector of 1 of size η

(iii) Also, $\mathbf{X}_i \in \{0, 1\}, \forall i$, is a constraint that needs to be satisfied for binary value of vector elements.

Therefore from Eqs. C.3, C.4, C.5 and C.6 we get,
Maximize:
$\mathbf{X}_0^T.\mathbf{X}_1 + \mathbf{X}_1^T.\mathbf{X}_2 + \mathbf{X}_2^T.\mathbf{X}_3 + \mathbf{X}_3^T.\mathbf{X}_4 + \ldots \mathbf{X}_\eta - 1^T.\mathbf{X}_\eta$
Subject to:
$\mathbf{BV}.\mathbf{X}_0 + \mathbf{BV}.\mathbf{X}_1 + \mathbf{BV}.\mathbf{X}_2 + \mathbf{BV}.\mathbf{X}_3 + \ldots \mathbf{BV}.\mathbf{X}_\eta > \omega$,
and $\mathbf{M}_0^T.\mathbf{X}_0 + \mathbf{M}_1^T.\mathbf{X}_1 + \mathbf{M}_2^T.\mathbf{X}_2 + \ldots \mathbf{M}_\eta^T.\mathbf{X}_\eta = \eta + 1$,
and $\mathbf{1}_\eta^T.\mathbf{X}_j = 1, \forall j \in \eta'$ and $\mathbf{X}_i \in \{0, 1\}, \forall i$
From Eq. C.1 we have

$$f(x) = x^t Q x = \sum_{i=1}^{n} \sum_{j=1}^{n} q_{i,j} x_i x_j, \quad x_i \in \{0, 1\}, \forall i$$

Subject to $a'x < b$

$$\text{Assume } q_{i,j} = \begin{cases} 1, & \text{if } i + 1 = j \\ 0, & \text{otherwise} \end{cases} \forall i \forall j$$

Even after this assumption the problem remains in the class of NP-hard [2],

$\Longrightarrow f(x) = \sum_{i=1}^{n} \sum_{j=1}^{n} q_{i,j} x_i x_j$
$\Longrightarrow f(x) = x_0x_1 + x_1x_2 + x_2x_3 + x_3x_4 + x_4x_5 + \ldots x_{n-1}x_n$
subject to $a'x < b$

This form of LCBQP problem represents the HTMS problem we have defined with variables replaced by vectors. Thus the LCBQP problem is reduced to the HTMS problem and hence lies in the domain of NP-hard problems.

References

1. Garey, M. R., & Johnson, D. S. (1990). *Computers and Intractability; A Guide to the Theory of NP-Completeness*. New York, NY, USA: W. H. Freeman & Co.
2. Merz, P., Freisleben, B. (1999) Genetic algorithms for binary quadratic programming. In *GECCO-1999: Proceedings of the Genetic and Evolutionary Computation Conference*, pp. 417–424. Morgan Kauffman.

Appendix D
Proof of Polynomial Bound Convergence of Algorithm 7.1

Lemma D.1 *Algorithm 7.1 converges in polynomial time for a node v, given that it has non-zero QoS metric (QM). In other words, the proposed algorithm has polynomial bound convergence time for any node which has sent/received any data traffic in the last 10 frames.*

Proof Consider an undirected graph $G = (V, E, L)$, with V, E and L as its vertices, edges between vertices and list of colors (containing C colors) available at each vertex respectively. For a node v under consideration with non-zero weights W in Eq. 7.7; any data traffic sent/received in last 10 frames will result in non-zero QoS metric (QM) for node v. And therefore the resource share calculated from it will be also non-zero. Algorithm 7.1 converges iff list CAL becomes empty. Therefore to prove Algorithm 7.1 converges it is necessary and sufficient to prove that all colors from list CAL_v are either allocated to node v or its neighbor $N(v)$ in polynomial time. For worst case scenario, the contention vector CV will never win the contention for selection of colors (lines 5–8 of Algorithm 7.1). But for every non-winning contention of node v, some other neighbor node w ($w \in N(v)$) may win the contention. If no nodes in $N(v)$ win the contention, then neighbor nodes of w i.e. N(N(v)) and so on, will eventually win the contention. And therefore, every winning contention in neighbor nodes will remove the colors from CAL. Since the algorithm is distributed in nature and all nodes executes concurrently, the maximum iterations by which all colors from list CAL are removed is determined by the diameter of the graph. In the worst case scenario, the maximum diameter is attained by a Path Graph with the diameter of $N - 1$ [1]. Therefore the worst case convergence time of the algorithm is $N - 1$ iterations, and hence it has polynomial bounded convergence time.

Lemma D.2 *Algorithm 7.1 converges in polynomial time for a node v having zero QoS metric, given that for each color l available in CAL, node v has at least one neighbor w with non-zero QoS metric and $l \in CAL$. In other words, if a node v has zero value for QoS metric (QM), it can still converge in polynomial time given that it shares common spectral opportunity with at least one neighbor with non-zero QoS metric for every color.*

V. Mishra et al., *QoS and Energy Management in Cognitive Radio Network*, Signals and Communication Technology, DOI 10.1007/978-3-319-45860-1

Proof Consider an undirected graph $G = (V, E, L)$, with V, E and L as its vertices, edges between vertices and list of colors (containing C colors) available at each vertex respectively. For a node v under consideration with non-zero weights W in Eq. 7.7; if no data traffic is sent/received in last 10 frames, it will result in zero QoS metric (QM) for node v. But since, node v shares common spectral opportunity with at least one neighbor with non-zero QoS metric for every color, the resource share (Eq. 7.8) is zero instead of non-defined. Thus the calculation of contention vector (line 5 of Algorithm 7.1) is a defined value. Since resource share is zero, the value of contention vector generally selected is large and hence has a low probability of winning the contention. In worst case scenario, the contention vector CV will never win the contention for selection of colors (lines 5–8 of Algorithm 7.1). This has been reduced to the proof of Lemma D.1, where it has been shown that the Algorithm 7.1 converges for worst case scenario where contention vector never wins the contention. Therefore this algorithm has polynomial bound convergence time.

Reference

1. Caccetta, L., & Smyth, W. (1992). Graphs of maximum diameter. *Discrete Mathematics*, *102*(2), 121–141.

MIX
Papier aus verantwortungsvollen Quellen
Paper from responsible sources
FSC® C105338

If you have any concerns about our products,
you can contact us on
ProductSafety@springernature.com

In case Publisher is established outside the EU,
the EU authorized representative is:
Springer Nature Customer Service Center GmbH
Europaplatz 3, 69115 Heidelberg, Germany

Printed by Libri Plureos GmbH
in Hamburg, Germany